Praise for *Vagina O*

An NPR and *Science News* Best Science Book of the Summer
A *Wired* Book You Need to Read
A *Book Riot* Best Book Out This Week

"[Rachel E.] Gross takes on a herculean task, exploring female anatomy from a medical, social and historical perspective. . . . Gross . . . makes palatable the sawing of cadavers and the injecting of silicone into two-pronged snake vaginas, without undercutting the gravity of their resulting revelations."   —Maya Salam, *New York Times*

"*Vagina Obscura* should cause a revolution in how we think about the vagina."   —Rosamund Urwin, *Sunday Times*

"This book is what we've been waiting for."
                                        —Francesca Brown, *Stylist* (UK)

"Gross describes . . . [how] better understand[ing] the female body adds to the growing understanding of gender and sex, and the illustrations by Veve serve to add wonder and beauty to the journey."
                                        —Erica Ezeifedi, *Book Riot*

"*Vagina Obscura* reinforces that female bodies are more than 'walking wombs' or 'baby machines.' Understanding these organs and tissues is important for keeping the people who have them healthy. It will take a lot of vagina studies to overcome centuries of neglect, Gross writes. But the book provides a glimpse into what is possible when researchers (finally) pay attention."   —Erin Garcia de Jesús, *Science News*

"The topic of this book is relevant to everybody."
                                        —Riley Black, NPR's *Science Friday*

"*Vagina Obscura* highlights how little most people know about vaginas. . . . Gross chronicles how a long history of medical misinformation has contributed to long-standing misunderstandings about reproductive anatomy."   —Laken Brooks, *Forbes*

"Enthralling and scrupulously researched. . . . Examining the vagina through historical, scientific, and political lenses, Gross crafts a piquant narrative that is exemplary of quality science journalism and is nearly impossible to set down. . . . A must-read."
                              —*Library Journal*, starred review

"[I am] just blown away by it. . . . The digging [Gross does], the folks [she] found, [her] ability to explain and keep the reader engaged. Great Science Writing."   —Mary Roach, on Twitter

"Through her seamless storytelling and meticulous research, Rachel E. Gross shows how long we have misunderstood the bodies of half the people who have ever lived. . . . *Vagina Obscura* is science writing at its finest—revelatory, wry, consequential, necessary, and incredibly hard to put down."—Ed Yong, Pulitzer Prize–winning journalist and author of *I Contain Multitudes*

"A marvel of a book—lyrical, compassionate, infuriating, insightful, and wise. Gross's exploration of the history, science, and politics of female anatomy should be read by women, men, and everybody seeking to be smarter about who we really are."
                              —Deborah Blum, author of *The Poison Squad*

"Mesmerizing and often humorous. . . . Rachel E. Gross brilliantly investigates questions regarding sex, sexuality, and reproduction that have been hidden, made shameful, or just never asked. "
                              —Elizabeth Reis, author of *Bodies in Doubt*

"With the perfect mix of verve and nerve, clarity and composure, Rachel E. Gross finally gives this organ system its due."

—Florence Williams, author of *Breasts*

"Well-researched and beautifully told. . . . A gripping read that should inspire significant change in science and society."

—Gabrielle Jackson, author of *Pain and Prejudice*

"The vagina is having a much-belated moment, and thanks to Rachel E. Gross, now so are the ovaries, clitoris, and uterus."

—Emily Willingham, PhD, author of *Phallacy*

"A lively debut with a fresh, informative examination of women's entire reproductive system, melding medical history—beginning in Hippocrates' Greece—with a wide range of interviews and biological sleuthing in research laboratories all over the world. . . . An eye-opening biological journey." —*Kirkus Reviews*

"As Rachel E. Gross notes near the beginning of her highly engaging book, 'You can't image what you can't see, but you can't see what you can't imagine.' Fortunately for us, Gross has a vivid imagination, one she draws richly upon as she shares her tour of the female genitals, enabling us to also see her discoveries."

—Sarah B. Rodriguez, historian and author of *The Love Surgeon: A Story of Trust, Harm, and the Limits of Medical Regulation*

"Finally, a guidebook to the vagina that does true justice to its complexity, both in its reality and in the cultural imagination. Beautifully written and incredibly detailed."

—Angela Saini, author of *Inferior* and *Superior*

"Gross writes with curiosity, humour, and often understandable anger."

—*Lancet* (UK)

Vagina Obscura

Vagina Obscura

AN ANATOMICAL VOYAGE

RACHEL E. GROSS

With Illustrations by Armando Veve

W. W. NORTON & COMPANY
Celebrating a Century of Independent Publishing

Copyright © 2022 by Rachel E. Gross
Illustrations © 2022 by Armando Veve

For information about permission to reproduce selections from this book, write to
Permissions, W. W. Norton & Company, Inc., 500 Fifth Avenue, New York, NY 10110

For information about special discounts for bulk purchases, please contact
W. W. Norton Special Sales at specialsales@wwnorton.com or 800-233-4830

Manufacturing by Lakeside Book Company
Book design by Chris Welch
Production manager: Lauren Abbate

Library of Congress Cataloging-in-Publication Data

Names: Gross, Rachel E., author.
Title: Vagina obscura : an anatomical voyage / Rachel E. Gross.
Description: First edition. | New York, NY : W. W. Norton & Company, [2022] |
Includes bibliographical references and index.
Identifiers: LCCN 2021049425 | ISBN 9781324006312 (hardback) |
ISBN 9781324006329 (epub)
Subjects: LCSH: Vagina. | Generative organs, Female
Classification: LCC RG268 .G76 2022 | DDC 618.1/5—dc23/eng/20211102
LC record available at https://lccn.loc.gov/2021049425

ISBN 978-1-324-05053-7 pbk.

W. W. Norton & Company, Inc., 500 Fifth Avenue, New York, N.Y. 10110
www.wwnorton.com

W. W. Norton & Company Ltd., 15 Carlisle Street, London W1D 3BS

2 3 4 5 6 7 8 9 0

This book was written for any woman—any person—who has found themself mystified by their own body. Anyone who has felt the nagging suspicion that what they have read about their anatomy wasn't written for them, or by someone like them. You were right. This book was written for you. It was written for anyone who has felt unable to talk about their body in language that others could understand. Anyone who wants to better understand the legacy they were born with, by virtue of their reproductive organs and the chromosomes dancing within their every cell.

I hope this book welcomes you in as the remarkable women and explorers in this book welcomed me.

CONTENTS

NAMED, CLAIMED, AND SHAMED

There comes a time in every woman's life when her body bumps up against the limits of human knowledge. In that moment, she sees herself as medicine has seen her: a mystery. An enigma. A black box that, for some reason, no one has managed to get inside. Those that I talked to for this book were all made to feel that they alone had a complicated, unruly body. They began to suspect, or were outright told, that somehow it was their fault. That they should be ashamed—should lie down and think about what they'd done.

My own moment came in July 2018. I was twenty-nine years old, and I had an itch I couldn't scratch. For the past month my vulva had felt on the verge of bursting into flames. At first my gynecologist, a tiny, no-nonsense woman named Dr. Lori Picco, thought I had a particularly exuberant yeast infection. But one antifungal treatment and two rounds of antibiotics later, she had bad news: my tormentor was a bacterial infection I'd never heard of. For half of women who get it, it pops back up again and again, without warning, like a Whac-A-Mole.

There was one last treatment I could try. "It's basically rat poison," said Dr. Picco. "You're going to see that on the Internet, so I might as well tell you now." It was called boric acid, and thanks to its fungi- and bacteria-killing powers, it's been used since the 1800s in antibacterial ointments, douching washes, and roach and ant killers. The idea was to nuke the ecosystem inside my vagina, wiping out the bad bacteria along with the good, so I could start afresh.

That didn't sound great. But neither did having a lifelong infection.

More than most, I was primed to take my doctor's suggestion. I was the daughter of scientists, a medical doctor (my mother); a theoretical physicist (my father); and a molecular geneticist (my stepmother). I had a master's degree in science journalism and worked as the digital science editor at *Smithsonian* magazine, where I fancied myself fluent, or at least conversational, in things like cells, biology, and bodies. To a large extent, I trusted medicine. I certainly trusted Dr. Picco, who had always approached my vagina with a curt, businesslike efficiency.

I took my poison like a good patient, every night, on my back. For ten days. But one night, I made a mistake. Exhausted from weeks of surreptitious itching, thinking about itching, and trying not to itch, I passed out early. When I woke up again, it was three a.m. and I had the tingling feeling that something was amiss. I stumbled to the bathroom, half-awake, and unscrewed my orange pill container. Then, without thinking, I swallowed my vagina poison.

I sat down on the toilet, hard.

I took out my phone and typed frantically into the Google search bar. The top result was a study entitled: "Fatal ingestion of boric acid in an adult." "Boric acid is a dangerous poison," it began. "Poisoning from this chemical can be acute or chronic."

I ran back to the bedroom and shook my boyfriend awake. But the words wouldn't come. Until that moment, I hadn't told him about the medication I was taking. Logically, I knew that whatever was happening to me down there had nothing to do with my self-worth. But in the back of my head, I still felt dirty. Contaminated. Radioactive. Like there was a force field of shame around my nether regions. Even for me, even in 2018, you just didn't talk about your burning bush.

"I think I swallowed something I shouldn't have," I said, my voice a child's whisper. He took one look at my phone and started putting on his shoes. It was time to go to the emergency room.

On the hospital bed, under the fluorescent lights, I remember feeling profoundly alienated from my body. I pictured myself getting my stomach pumped, convulsed by shockwaves, taken over by a force greater than myself.

Beneath that disconnect was something else. Betrayal. Outrage. This wasn't supposed to happen. I thought of myself as an educated, rational, science-minded woman, equipped with the tools to control my own life. Why didn't I know the inner workings of my own body? Why didn't my gynecologist, or any of the other doctors I'd encountered—medical experts who had spent their lives studying and caring for bodies like mine? Why, for God's sake, had my gynecologist recommended I shove rat poison up my vagina?

It hit me: I knew almost nothing about how my vagina worked.

That was the moment this book was born.

I set out to write a book on the science of vaginas. It would be fun and jaunty and full of wonder, a Ms. Frizzle–esque journey into the intimate depths of the human body. But I soon realized I had a problem: there is a vast knowledge gap when it comes to what we know about the female body. Most of our scientific understanding of this realm is built off of the study of male bodies. It was only in 1993, following the women's health movement, that a federal mandate required researchers to include women and minorities in clinical research. As Dr. Janine Austin Clayton, associate director for women's health research at the

National Institutes of Health, put it in 2014: "We literally know less about every aspect of female biology compared to male biology."

Until recently, medical research on women in this country was centered mainly on fertility. As I was told by one endometriosis expert: "Nobody in Congress really cares about the uterus when it doesn't have a baby in it." Only in 2014 did the NIH start a gynecological branch to look at the health of vulvas, vaginas, ovaries, and uteruses in their own right. That was virtually the first time a federal research branch acknowledged that these organs are integral to women's health, whether or not they intend to get pregnant. (Even that research, however, is housed within the National Institute of Child Health and Development. "What part of NIH is interested in lady parts? It's not intuitive," says Dr. Diana Bianchi, the institute's director. "But in fact it is largely us.")

As a result, there are parts of your own body less known than the bottom of the ocean, or the surface of Mars. Most researchers I talked to blamed this dearth of knowledge on the black-box problem: the female body is considered more complex, more obscure, with much of its plumbing tucked up inside. To get inside it, we've needed high-tech imaging tools, tools that have only come around in recent decades. When I heard these answers, I couldn't help thinking of what science has done in the twenty-first century: put a rover on Mars, made a three-parent baby, built an artificial sheep uterus. And we couldn't figure out the composition of vaginal mucus?

It wasn't just a lack of tools, I would learn. It was a lack of will. Going back to Darwin, scientists have considered vaginas and their accoutrements less interesting, less important, and less dynamic than penises. They either didn't care, or felt embarrassed, or insisted on thinking of women as only reproductive, not sexual. "You can only see what you're looking for," I was told by Scott Pitnick, a sperm biologist who studies the interactions between sperm and female fluids. "If you're not expecting females to be important or to make a real contribution, you're just not going out and actively studying them."

When pressed, he and others acknowledged that this discrepancy, at its heart, is due to sexism within science and who is actually doing the investigating. For most of history, women—especially women of color, trans women, and women who are sexual minorities—have been excluded from this supposedly universal endeavor. As I reported this book, it became clear to me that the two problems were inseparable. The marginalization of women's bodies from science is largely due the marginalization of women from science.

⌐

You can trace these biases to the birth of Western medicine, to the OG doctor himself, Hippocrates. That most revered of Greek physicians, namesake of the Hippocratic Oath, never actually studied a human woman. Due to cultural taboos and a dearth of female corpses, he relied mainly on the words of midwives and women who performed self-examinations. "I only know what women have taught me," he said. That detail did not stop him from naming our sexy bits.

Around 400 BCE, Hippocrates deemed the male and female genitalia τὸ αἰδοῖον. It means "the shame parts" and is a reference to Aidos, blushing goddess of shame and modesty. The study of sexuality, in other words, was tinged in shame from the beginning. But it was women to whom the shame stuck. In 1545, a French anatomist dissected a human clitoris and named it *membre honteux:* the shameful member. Many medical textbooks still refer to the vulva—all of the outer female genitalia—by the Latin word *pudendum*, which translates into "part for which you should be ashamed."* The German word for labia is *Scham-lippen*—"shame lips."

These historical roots radiate out to today. When I told people what I was writing, I was surprised by how eager many were for a book like this. When I told my colleague, Beth—a brash, wickedly funny editor

---

* Although, in 2020, the term was officially retired from the international dictionary of anatomical terms after heated debate.

at *Smithsonian*—she looked at me with a kind of awe. For the two years we'd worked together, Beth had kept a hand-sewn ovary made of lavender felt on her desk, a memento of the flesh-and-blood gonad she'd had cut out after being treated for reproductive cancer. Yet even at fifty-five years old, she admitted, she couldn't bring herself to say the word "vagina."

She leaned close and told me with embarrassment what her parents had taught her to say instead: "front butt."*

Beth is not alone. We are a society profoundly unable to talk about our own anatomy—out of both ignorance and an ancient sense of shame. Surveys find that nearly half of British women (and a quarter of American women) can't identify the vagina on a medical diagram; fewer know where the vulva or cervix is. Children in England are taught to refer to their vaginas using cutesy words like "tuppence," "foo-foo," "fairy," "fanny," or "minnie." How can we expect those children to grow into adults who are knowledgeable and authoritative about their own bodies, if they don't have the language to discuss them?

These moments made me sad, outraged, and confused. They reinforced why I was spending three years of my life calling up people to ask about vaginas: young vaginas, aging vaginas, duck vaginas, dolphin vaginas, vaginas in health, vaginas in sickness, vaginas in flagrante, and bioengineered vaginas built in labs. The reason I was doing it was that no one else had. And we were all suffering the repercussions—whether we knew it or not.

———

The consequences of my own lack of self-knowledge were mild. After the ER doctor called Poison Control, he reassured me that the dose I'd taken wasn't nearly enough to necessitate a stomach-pumping. Not

---

* Beth now proudly uses the word "vagina" in everyday conversation, even when it isn't strictly necessary.

long after, my infection cleared up. But like the other women in this book, the experience started me on a journey to greater understanding. Many of them also found themselves in hospitals, in experimental research trials, or on the front lines of reproductive research, searching for answers to intimate questions that affect us all. For them, the consequences were often orders of magnitude more severe.

Today, women and LGBTQ scientists are surveying this realm anew, and uncovering what generations before them have missed. The landscape they are charting may as well be a different planet from the shame-filled world that Hippocrates imagined: a world where the clitoris reigns over its vast subterranean kingdom; the vaginal canal teems with bacterial soldiers; and the ovaries revitalize themselves by pumping out brand-new eggs.

This book will take you to the anatomical labs of the University of Melbourne, to meet the doctor who was taught in medical school that the clitoris was small and diminutive, and who now uses modern imaging techniques to reveal the true shape of this majestic organ (Chapter 2). It will take you to the university where a Boston biologist is creating artificial ovaries in the hopes of protecting her daughter's generation from some of the cascading health effects associated with menopause (Chapter 6). It will take you to San Mateo, California, to meet the surgeon who is transforming gender-affirmation surgery, so she can give patients the experience she wishes she'd had when she had her own surgery in Mexico in the 1990s (Chapter 8).

It will also take you to the annals of history, where you'll find that women have always been a part of this story—often behind the scenes, without credit, and outside of the official academy. We'll make stops in Hippocrates's ancient Greece, Darwin's Victorian England, and Freud's interwar Austria—touchstones in the traditional narrative of how the female body was won. These men saw themselves as bold explorers planting flags in uncharted territory. But they were not the first to go here—nor would they be the last. Beyond those well-worn tales, you'll meet dogged bench scientists like Miriam Menkin, the first researcher to fertilize a human egg outside the body (Chapter 5), and trailblazers like Princess Marie Bonaparte, who, despite her lack of formal medical training, uncovered new data about the clitoris at a time when that organ was dismissed and despised (Chapter 1).

Today, even the concept of the female body is undergoing a profound transformation. Science has long separated the vast sweep of human bodies into two boxes: male and female. Much of modern medicine is predicated on the assumption that sex can only be either/or, two trains that run along separate, parallel tracks. But again and again, biology has proven this not to be the case. We now know that sex and gender are not binary, and that identity, chromosomes, genitalia, gonads, and hormones rarely form a neat line. Bodies exist on a spectrum, a permutation of endless forms most beautiful.

The more we embrace these connections, the more we advance the science of all bodies. Researchers who study endometriosis, for example, find that patterns of inflammation that underlie this disease also affect the health and fertility of male bodies. New research into the vaginal microbiome helps us better understand the role of the penile microbiome. And studying the regenerative powers of the testes lets us see the ovaries anew, not as dwindling reserves but as hotbeds of growth and regeneration.

———

One of the questions I found myself asking scientists most often as I reported on this book was: why has it taken until now for science to investigate [insert obvious thing]? For example: What makes a healthy vaginal ecosystem? How does the menstrual cycle actually work? What is the G-spot, really? In response, I always heard some version of the phrase: You can't see what you aren't looking for. Or: you see what you expect to see. In many ways, this book is about different ways of looking.

Science is done by scientists. They live in their own eras, in their own skin. They look at the world not only through microscopes and telescopes but through their own, limited, human lens. And for most of human history, these scientists have been Western, white, and male. They were shaped by the attitudes and politics of their time—and the knowledge they produced in turn reinforced and perpetuated those politics. Throughout history, this scientific knowledge has been used to silence some and privilege others, to decide which bodies were worthy and which were not.

I hope this book can illuminate the blinders that limited what those early anatomists saw. To challenge the idea that what they produced was objective knowledge. To show how beyond that horizon, there was more to see and know.

When these men looked at women, they often peered through a lens of reproduction: women were walking wombs, baby machines, sexual difference. Today, a new generation is thinking outside this box. They're looking at those organs most bound up in reproduction—the uterus, ovaries, vagina—and seeing them as part of a larger whole; as being dynamic, active, and resilient; as a window into more universal processes like healing and regeneration.

You can't imagine what you can't see, but you can't see what you can't imagine. The people and discoveries in this book are a testament to what we can see, if we imagine differently.

ON THE INADEQUACY OF LANGUAGE

In writing this book, I realized there is no good word for what we're talking about: The constellation of organs that take part in sex, birth, reproduction, and so much more. In medical parlance, they're referred to collectively as the "female reproductive system." But that phrase, besides being a mouthful, doesn't really cover it. The clitoris, for instance, often isn't considered a reproductive organ at all. Meanwhile, there are reproductive organs I don't cover, like the breasts. (Or the bone marrow, which some researchers argue should be considered reproductive, because stem cells from bone travel to the uterus during the menstrual cycle to help the uterine lining regrow.)

There are other reasons why "reproductive system" doesn't get the job done. The ovaries don't just grow egg cells; they pump out a hormone cocktail that maintains the health of virtually every organ system, from heart to bones to brain. The vaginal microbiome is an extension of your body's immune system, protecting this liminal space from intruders while helping keep the body's equilibrium in check. The uterus is one part of an intricate, body-wide dance, trading in stem cells and immune cells with the blood and bone marrow. Together, these organs make up something larger, a web of rivers and pathways that feed into each other and work together to keep the balance of you.

The lack of a good term for such an integral group of organs is telling. This is a region of the body we still find hard to talk about, both literally and figuratively. Silence, stigma, and, yes, shame still stunt the conversation—and the science—about the female body. Ultimately, I used the word "vagina" in the title, because it has the name recognition. But know that while this book started out as a book on the science of vaginas, it is also about so much more. This is a book about why we don't have a good name for the vagina et al.—and maybe, a glimpse into a future where we do.

———

Throughout this book, I use the word "woman" in a few different ways. Often I use it in a historical sense, to show how men have drawn a line around those with certain body parts and slotted them into the category of "woman." To these men, being born a woman meant being destined to be a mother, a wife, a helpmate, "a little creature without a penis," as Sigmund Freud charmingly put it. I think it is important to illuminate the anatomical criteria they used to create these categories, which would later be used to constrain and constrict the people who had those bodies.

In reality, not all people with vaginas and uteruses are women, just as not all people with penises and testicles are men. Even objective-sounding terms like "sex," "male," and "female" are not pure, and none represent a clear biological choice. There are women born with larger-than-usual clitorises, or without ovaries, or with elevated testosterone levels. There are women who lose their ovaries and uteruses to hysterectomies or other medical interventions. There are also people who don't fall under the umbrella of women, but who do have the body parts I'm talking about, and who throughout history have been judged accordingly: intersex people, nonbinary folks, trans men. I hope they, too, will find something of value here.

In this book, I show some of the ways that science has defined woman in the past. In my efforts to write this history, I will sometimes fail in my effort to help us move past their binary definitions. But ultimately, I hope we—you, me, science—can redefine her. Let me be clear: a woman's essence is not in her womb. She is not defined by her biology, society, science, men, or other women. She is defined by her knowledge that she is a woman. It is you, reader, who decides whether "woman" refers to you—or not.

Would it not for this they would have no desire nor delight.

—BRITISH MIDWIFE JANE SHARP, 1671

CHAPTER I

Desire

(GLANS CLITORIS)

The princess lies supine on a hospital bed, waiting for her life to begin. Her arms are folded over her chest, her eyelids lowered, as if in slumber.

It is spring 1927, and Marie Bonaparte has traveled by train from her Parisian chateau to the Löw Sanatorium, a private clinic in the heart of Vienna. She is about to undertake the most dramatic decision of her life. The forty-four-year-old princess has left behind her prince, Prince George of Greece and Denmark, and their two teenage children to get an experimental operation developed by one Dr. Josef von Halban. She knows this operation well because, in 1924, she wrote an article advocating for it in a medical journal. In her description of it, "the suspensory ligament of the clitoris (is) severed and the clitoris secured to the underlying structures, thus fixing it in a lower position."

Marie was desperate. Twenty years earlier, newlywed and in her prime, she learned something that would make most women weep: she was frigid. In her interpretation of the word, that meant she couldn't reach orgasm in the missionary position. She had certainly tried. Soon after marrying the tall, blond, mustachioed prince of her dreams, she learned that he was romantically attached to his uncle and had little sexual interest in her. "I hate it as much as you do," George told her on their wedding night. "But we must do it if we want children." She did her duty, giving birth to two healthy heirs, a boy and a girl. Then she embarked on a series of passionate affairs with high-ranking men across the Continent, including Aristide Briand, the eleven-times prime minister of France.

Marie was a handsome woman, with long brown curls, a knowing half-smile, and a pensive, heart-shaped face. Outwardly, she looked like a paradigm of queenliness and maternity. Inwardly, she was a woman divided. None of her affairs brought her fulfillment. "My libido is all in my head," she confided in her journal. She couldn't know, of course, that the coveted "vaginal orgasm" was not exactly a routine occurrence. All she knew was that her body refused to obey her mind.

The princess's problem went beyond the bedroom. Under the reign of her great-granduncle Napoleon, France had become "la Grande Nation," a vast colonial empire with a booming population. Now, after the First World War, a depleted France was desperate to replenish its lost generation of fallen men and reverse the plummeting birthrate—a burden that fell, naturally, on women. Doing one's patriotic duty meant bearing more children. And bearing more children meant enjoying matrimonial coitus. Thus the mark of this ideal Frenchwoman—"normal, vaginal, maternal," as Marie put it—was that she took pleasure in penetration. Coming during sex was neither frill nor perk, but social necessity.*

Try as she might, Marie found herself unable to obey this directive. "While work is easy, pleasure (*la volupté*) is difficult," she would write. Her years of psychoanalysis had done nothing to help her problem. Now, she had decided to find a different way. She would move her clitoris southward, so that she could finally feel pleasure during inter-

---

* Not just in France: In 1920s America, a woman failing to orgasm during sex with her husband was legal grounds for divorce.

course. Only with this surgery, she believed, would she achieve the erotic harmony she desired.

In the operating room, Halban lifted Marie's hospital gown to expose her vulva. He injected her with local anesthesia; she would be awake during the procedure. Marie's pelvis grew numb. She was not nervous. In a way, her whole life had been leading up to this. Her American friend and fellow budding psychoanalyst, Ruth Mack Brunswick, sat beside her. Ruth was fifteen years younger, but the two were so close that they shared masturbation techniques. (Marie wrote that Ruth was "prouder of her masturbation than of ten doctoral degrees.") In their letters, they discussed their own patient cases of frigidity and "clitoridism," Marie's self-diagnosis.

Halban took a scalpel and carefully began cutting into the delicate skin above Marie's glans clitoris, the pea-sized nub that sat just above the apex of her vagina. Then he sliced through the ligament that anchored her clitoris to the pubic symphysis, the band of tough cartilage that seals the two halves of the bony pelvis together. Having freed it from its constraints, he now lifted her clitoris—her jewel, her phallus, the seat of her pleasure—and moved it a few millimeters down, so that it sat nearer to her vaginal opening. When he was satisfied, Halban sewed it back on. He then sewed up the wound where Marie's clitoris once was, to minimize scarring. The entire operation took just twenty-two minutes.

One can only imagine the pain Marie found herself in as the anesthesia wore off. She stayed in the hospital for two weeks, with Ruth at her bedside. But perhaps the most important figure in her life was missing. Where was Freud?

———

Sigmund Freud ignited something in Marie. She had read the French translation of his *Introductory Lectures on Psycho-Analysis* in 1923, long before clitoral surgery was even a twinkle in her eye. At the time she was caring for her father, who was suffering from prostate cancer. After his death, trapped in her childhood home, she began rereading the journals she had written between the ages of seven and ten—a dark tangle of poems and stories that told of a paranoid, overprotected childhood. She became enamored with the budding field of psychoanalysis as a way to reinterpret her life. Marie sensed in Freud an answer to her questions—why her romantic entanglements never seemed to fulfill her, why themes from her childhood kept resurrecting themselves in dreams.

She decided she would meet this revered architect of the unconscious. He was, she would later write, the only one who could help her find "the penis and orgastic normality" she was seeking.

Freud, for his part, was not exactly thrilled at the prospect of meeting Marie. He was an old man, nearly seventy, when a mutual friend reached out to him about her case. Freud was revered, to be sure—his Vienna Psychoanalytic Society had just put down roots in the United States and Russia—but also deep in melancholy. He had lost his beloved daughter Sophie to Spanish flu four years earlier, and her son, his grandson, soon after. Now, after an operation on his jaw to remove a malignant tumor, he wore an unwieldy prosthetic that made talking painful and smoking difficult. Isolated by age, consumed by grief, he was taking few new patients, and was extremely picky about whom he chose. Before Marie had come to him, he would later tell her, he had almost lost his will to live.

The mutual friend, an Alsatian psychoanalyst named Dr. René Laforgue, vouched for Marie, telling Freud she was "sensible and conscientious." Marie "has, in my opinion, a marked virility complex," Laforgue wrote, adding that she was interested in analysis "for didactic reasons"—to learn.

The prospect of a patient who wanted both to be cured and to learn analytic techniques was unusual. Marie's special requests also raised Freud's eyebrows. For one thing, she wanted him to see her twice a day, for a total of two hours. "With the princess, it looks as if nothing can be done," he wrote back to Laforgue. "Since I take only very few cases, an analysis of six to eight weeks, forcing me to give up another one and extending over a season, cannot tempt me." But when Marie wrote directly to Freud, taking the time to translate her first draft from French into German, he relented. "My venerable master," her letter began. On September 30, 1925, Marie visited Freud.

Freud had lived and worked for more than thirty years at Berggasse 19, in a middle-class, residential neighborhood of Vienna. The five-story building stood on a cobblestone street within walking distance of the university and the Psychoanalytic Institute. On the street level were a butcher's shop and a food co-operative. On the second floor, Freud's famous study, with its squashy, velvet-draped couch and its bookshelves piled high with psychoanalytic texts, looked out over the backyard. Adjacent to it was the Freud family apartment, where he and his wife, Martha, had raised their six children. To reach Freud's study, Marie entered the stone archway from the street and climbed a wide set of stairs from the foyer, the same stairs hundreds of patients had climbed before her.

The two hit it off. Marie, for Freud, was a breath of youth and fresh air. An aristocratic woman who had rubbed shoulders with great minds of art and science, she proved an eager student of his theories. She fawned over the father of psychoanalysis, giving Freud the appreciation he needed and the adoration he craved. For Marie's part, Freud filled the role of father figure and mentor. In less than a month, she was referring to him as "my dear friend" and even saying she loved him. (When she returned home, in what can only be described as a Freudian slip, she accidentally left her wedding ring behind in her Viennese hotel.) Soon, Freud confided in Marie his journey with cancer, his grief, his money woes. He called her *meine liebe prinzessin*—my dear princess. She called him *Maître Aimée:* "beloved master."

"What a marvelous, unique being," Marie wrote in her journal. "Daily contact with such a mind is the greatest event in my life."

Freud thought he understood Marie's problem. Well-adapted women only desired vaginal penetration. Marie did not. This was because she had never fully accepted her role as a woman. She was a perfect example of what he called the "female masculinity complex": a woman who had failed to reconcile her Oedipal wish to replace her father. Instead, Marie had become obsessed with her clitoris—or her female phallus, as Freud liked to call it. He noted that Marie was drawn to "phallic women." One in particular was her late grandmother, Princess Pierre Bonaparte, widely known for her mastery of horse-riding, boar shooting, and upright pissing, which she accomplished by squatting and pulling her skirts down.

"My grandmother, tall, strong, and severe, with her imperious voice and single, steady black eye, and a few hairs on her chin, represented perfectly a 'phallic woman,'" Marie later wrote. In her mind, Princess Pierre threatened "to deprive little girls of their phallus-clitoris as a punishment for their sexual sins, while keeping an imposing phallus for herself."

Freud decreed that Marie was "bisexual"—not just in the way he believed all humans were bisexual, retaining "slight remnants of the stunted sex," but in that she had a pronounced male character. This was a compliment: he had come to associate bisexuality in women with heightened creativity and achievement. To her delight, he also declared that she had "no prudishness whatsoever" and that other female analysts had "neither your virility, nor your sincerity, nor your style." Freud's pronouncements, and his growing interest, pleased Marie. She soon became his favorite, drawing the envy of even his daughter Anna. He relented to her request of two hours a day and allowed her to take copious notes during her analysis. "Nobody understands you better than I," he told her.

But his understanding of her, like his understanding of female sexuality, was incomplete.

Marie's time with Freud brought her "peace of mind, of heart, and the possibility of working," she wrote, "but from the physiological point of view nothing." After a year of her analysis, she had still not found satisfaction. On July 29, 1926, she had her first consultation with Dr. Halban.

Marie was no stranger to surgical fixes. She'd had procedures to "correct" her breasts, smooth out a scar on her nose, and remove an ovarian cyst. But she knew that, this time, surgery represented a conclusion of sorts. It was the "end of the honeymoon with analysis," she wrote in her journal, an admission that psychoanalysis was not enough. It was the culmination of years of her own research, and a growing divergence between Freud and her when it came to female sexual development. For too long, the pinnacle of femininity—the vaginal orgasm—had mocked her with its absence. In 1927, she went under the knife.

Marie's dilemma was not unique to her time. The idea that the components of the female genitalia exist in constant conflict—that women have a vagina and a clitoris, and never the twain shall meet—has deep roots.*

In ancient Greece, the vagina reigned supreme. The reason goes back to how ancient thinkers thought about sexual difference. Greek physicians tended to consider men and women as essentially the same, a concept that historian Thomas Laqueur deemed "homology" or the "one-sex model." In this framework, the vagina (and sometimes the uterus) was the equivalent of an interior penis; the ovaries were interior testicles.

There is some anatomical truth to homology. At six weeks after conception, you were just a wriggling comma with limb buds, bob-

---

* Spoiler: so does the clitoris.

bing along in your mother's juices. But you were pulsing with potential. Between your legs was something called the genital tubercle, a fleshy knob with two parentheses-shaped swellings beneath it. That knob had the capacity to grow into either the clitoris or the penis; the swellings could become labia or scrotum. Between them was an opening that led to two pairs of ducts: the Müllerian or paramesonephric (female) ducts and the Wolffian or mesonephric (male) ducts. At eight or nine weeks, in a female fetus, the Wolffian pair of ducts wither away, and the Müllerian ducts fuse together like a wishbone to create the upper vagina and the uterus, with the tips becoming the Fallopian tubes.

Medical textbooks once attributed this divergence solely to the hormone testosterone, which is secreted by the fetal testes. In the absence of testes, the female body plan supposedly took over "by default." Femaleness was an iPhone on factory settings; maleness meant adding the baubles and whistles. Actually, a suite of active factors are required to trigger development of the ovary, an equally complex structure (more on that in Chapter 6). In a typical female, genes, hormones, and other factors push cells in the belly to become an ovary, which in turn helps tell the knob to grow into a clitoris, the swellings to form two sets of labia, and the ducts to form the uterus, tubes, and other structures. In a typical male, testosterone helps guide the growth of the penis and the fusing of the swellings into scrotum, which will eventually house the testes.

By ten weeks, the clitoris and the penis are roughly the same size. Only by the end of the third month do these two forms diverge, the penis protruding outward and the clitoris growing inward. By the end of the fourth month, in a female, both sets of labia and the clitoris are clearly defined. (That's why twelve weeks is the earliest a doctor can do a sex test by ultrasound.) So the Greeks weren't entirely off-base: It's just that, in reality, the equivalent of the penis is not the uterus or the vagina, but the clitoris. The scrotum, meanwhile, is homologous to the labia majora, while the testicles are indeed akin to the ovaries. These parallels don't go away: into adulthood, these organs share remarkably similar tissues, cells, and structures.*

Yet for the Greeks, homology went beyond biology and bodies. Anatomy reflected the divine order—an order in which women were forever inferior and imperfect. "Thus the male body was set as both standard and ideal; women were a lesser, stunted version," Laqueur wrote. "They have exactly the same organs as men, but in exactly the wrong places." In a practical sense, this meant women needed the same equipment: a penis, testicles, and the ability to make sperm. Thus Hippocrates declared that women produced female sperm, or "female essence," and that both partners must ejaculate during sex for conception to occur. On the plus side, this implied that women were capable of experiencing—and in fact must experience—great sexual pleasure.

Hippocrates (instructed by midwives, no doubt) accurately located and identified the clitoris. He called it *columella* or "little pillar," a name that may have acknowledged its capacity for erection. Not long after, Aristotle noticed something similar. In rats, a part of the female's pubic area swelled when she approached a male. He connected this finding to humans, noting that women "feel pleasure from being touched in the same place as men, but, in their case, there is no liquid emission." In the first century CE, another Greek physician, Soranos of Ephesus, described the position of the clitoris and may have given it its modern-day name: *kleitoris,* or nymph. "This small formation is called the nymph," he wrote, "because it is hidden underneath the labia such as young brides under their veil."†

---

* What's more, remnants of each sex remain in the other. Men have a tiny Y-shaped vestige of the paramesonephric ducts in their urethras, known as the prostatic utricle or *uterus masculinus*. In women, part of the mesonephric ducts persists as a structure called the appendix vesiculosa, a dangling body the size of a small pea that hangs from a stalk off the side of the Fallopian tube.

† By the seventeenth century, "nymphae" had become an umbrella term for the entire outer female genitalia, similar to the word vulva. Later, it came to specify the labia minora.

But it was Galen of Pergamon who solidified woman's sexual lot. This influential second-century physician took homology to its logical extreme. He envisioned the female reproductive system as an inverted penis, with the uterus a hollow phallus and ovaries as internal testicles. "Turn outward the woman's, turn inward, so to speak, and fold double the man's, and you will find the same in both in every respect," he wrote. Once, he memorably compared a woman's sexual organs to the eyes of a mole. He argued that both were best thought of as "unexpressed organs": internal, less functional, and less perfect than those of other animals. For Galen, only the vagina and its reproductive capacity mattered.

Galen's model left little room for the clitoris. He omitted it entirely from his anatomical descriptions—and with it, women's libido and capacity for sexual pleasure. Compared to what Hippocrates had written five hundred years earlier, this was a striking reversal. Unfortunately, Galen's was the school of thought that would hold sway until at least the seventeenth century, shaping anatomical ideas for centuries to come. Thus began a dark age for the clitoris. Forgotten by men of science, it languished for centuries, waiting for a prince to awaken it and rescue it from obscurity.

And finally, one did—well, actually, two.

———

In a temporary wooden theater, rings of concentric wooden circles held hordes of eager students, jostling to see the human body give up her secrets. Music played softly in the background. This scene might have taken place in a local church, or a courtyard. Either way, it could only have occurred in winter, to avoid the stench of rotting flesh wafting up on warm summer air. Onstage, the father of modern anatomy, Andreas Vesalius, began to slice into a corpse, usually a criminal who had been hanged. With a flourish, he displayed each body part to his spectators—flesh, muscle, bone. In one portrait, he stares at the viewer while revealing the flayed muscles and tendons of a human arm, just as another Renaissance scholar might display a book.

In sixteenth-century Italy, the body had become the new frontier of the mind. And once the human form became the canvas on which great explorers could pen their discoveries—well, then men started paying attention to the clitoris.

Vesalius began swiftly dismantling the framework Galen had erected, correcting hundreds of flaws in the old master's work. For one, he rejected Galen's description of the uterus as having horns and two chambers, noting that "Galen never inspected the uterus of a woman, unless it was in a dream, but only those of cows, goats, and sheep."* Yet Vesalius too suffered from a dearth of female corpses. Thanks to his reliance on hanged criminals, he is believed to have dissected very few women. He continued to rely on Galen's idea of the uterus as interior penis, illustrating it as a hollow penis with curly pubic hair at the tip. Perhaps this explains his odd conclusion that healthy women possessed no such thing as a clitoris—rather, he claimed this structure only existed in "hermaphrodites."

Matteo Realdo Colombo begged to differ. Once Vesalius's assistant, the younger anatomist would go on to become his mentor's greatest challenger. In his *De re Anatomica*, published posthumously in 1559, Colombo announced that he had discovered the clitoris.† Needless to say, this was not the case. But he did highlight its role in female pleasure. While Vesalius scoffed at "this new and useless part," Colombo poetically exalted it as "this so beautiful thing formed by so great art." He deemed it "the principal seat of women's enjoyment in intercourse, so that if you not only rub it with your penis, but even touch it with your little finger, the pleasure causes their seed to flow forth in

---

* These animals have bifurcated uteruses that are split down the middle, and curl outward like rams' horns.

† By this time the clitoris had been known to Greek, Persian, and Arabic writers for over a millennium, not to mention . . . women.

all directions." Like any self-styled explorer describing a new world, Colombo set out to mark his territory: "Since no one else has discerned these processes and their working; if it is permissible to give a name to things discovered by me, it should be called the love or sweetness of Venus (*amor veneris*)."

Two years later, an Italian anatomist named Gabriele Falloppio accused Colombo of treading on territory already claimed. In his 1561 book *Observationes Anatomicae*, he wrote that "this part is hidden and ignored by anatomists . . . it is so hidden that I was the first to discover it and if others speak of it, please know that they learnt about it from me and my students!" Today Falloppio is remembered not for his clitoral catfights but for naming the vagina and placenta, correcting the popular idea that the penis enters the uterus during sex, and bestowing his name upon the Fallopian tubes.

The seventeenth century saw a vast flowering of clitoral knowledge. In her 1671 birthing manual, Jane Sharp, a British midwife, described the clitoris as a small phallus that swells "when the spirits come into it" and "makes women lustful and take delight in copulation, and would it not for this they would have no desire nor delight." The following year, Dutch phsyician Regnier de Graaf wrote what is likely the first comprehensive account of female genital anatomy. He used the opportunity to chastise his colleagues for their failure to acknowledge the clitoris: "We are extremely surprised that some anatomists make no more mention of this part than if it did not exist at all in the universe of nature," he wrote. "In every cadaver we have so far dissected we have found it quite perceptible to sight and touch."

Soon a very different picture of this organ was emerging. In 1844, Georg Kobelt, a German anatomist and medical illustrator, dissected human reproductive organs, injected them with colored ink, and made intricate line drawings of what he saw. He noted how the shaft of the clitoris bent downward like a knee, and commented on the richness of nerves, "beautifully developed," in the glans. His images reveal two plump bulbs that arc down from the glans clitoris—the part often thought of as the entire organ—to encircle the vaginal walls. These bulbs appear to be made of a squiggly, brainlike substance that is actually erectile tissue, the same as that found along the shaft of the penis. The full clitoris, Kobelt wrote, "is comprised of two areas, the clitoris and the spongy bulbs . . . the two areas are connected by a vascular network." This was no external button but an extensive organ richly supplied with blood and nerves. *

Anatomists had definitively ousted the grim Galenic vision of female sexuality. The next logical step would have been to continue mapping this structure to realize that it encompassed far more of the female genitalia than anyone had suspected. Yet this was not the direction the arc of history was destined to bend. Instead, within decades the clitoris would be demonized, dismissed, and left to the trash heap of history. What happened? In a word: Freud.

———

Marie knew how her clitoris worked. As a young child, she had no qualms about touching herself where it made her feel good. But one day, her innocent pleasure was cut short.

Marie's nurse, Mimau—a nickname Marie had given her when she had appeared in a doorway holding a bouquet of bright mimosa flowers—slept in an adjoining room at the family's Saint-Cloud estate. The round-faced Corsican widow was the child's primary caretaker, after her mother died following a traumatic childbirth. Mimau, having no living children of her own, loved Marie dearly. But she was a devout Roman Catholic, and for her, masturbation was a sin that would lead to disease of Marie's body and soul. One night, she came into Marie's

---

* He also compared human clitorises to those of horses, cats, dogs, rats, pigs, rabbits, and lemurs, noting that these animals had a clitoris bone (*os clitoridis*) running through the middle. The males of those species have an equivalent penis bone (*baculum*).

room to find the young girl with her hand in her pajamas. She yelled: "It's a sin! It's a vice! If you do that, you will die!"

She soon forced Marie to wear a strange contraption: a nightgown with drawstrings at the bottom to prevent her from the forbidden activity.

Marie was already a morbid child. Her mother's abrupt passing had left her haunted with guilt and convinced that she, too, would die an early death. After a bout of tuberculosis, she grew terrified of germs, illness, and the morphine-laced cold medicine *sirop de Flon*. Mimau's scolding moved her deeply. Fearing that her bodily pleasure would lead to an early demise, she gave up masturbating when she was eight or nine years old.

Mimau was not alone in her masturbatory concerns. In the late nineteenth and early twentieth centuries, masturbation panic set Europe ablaze. For women, these anxieties settled on the clitoris—thanks in part to those same anatomists who had deemed it integral to female pleasure but useless to reproduction. Clitoral orgasm threatened the idea that married couples should be solely devoted to procreation. Along with short haircuts, cigarettes, and boxy clothes, it was unbecomingly masculine and not "in the interest of the reproductive health of the nation" writes historian of medicine Alison M. Downham Moore. Some doctors believed a visible clitoris to be a sign of "women who allowed themselves to be led astray by their passions, leading to neuroses such as hysteria (and) nymphomania that sometimes lead to death."

In a way, Marie was lucky. Besides "preventative undergarments," there are reports of masturbating children in the 1890s who were immobilized with braces and leg-separating contraptions, and even one whose hands were tied to a collar around her neck (her feet, inexplicably, were tied to the footboard). If such measures didn't work, doctors turned to surgery. In the United States doctors frequently removed the clitoral hood in an attempt to "liberate" the clitoris from perceived irritation. In the 1850s, an English gynecologist named Isaac Baker Brown recommended a full amputation of the clitoris—known as clitoridectomy, the surgical removal of the glans and shaft—as a cure for "nervous disorders" like hysteria, homosexuality, and epilepsy.* At least forty-eight women, ranging from sixteen to fifty-seven years of age, would lose their clitorises to his scalpel.

In 1866, Baker Brown was roundly denounced and removed from his post as president of the Medical Society of London. But it turned out other physicians didn't hate his surgery—they just thought it should be applied with more restraint. Doctors in the United States would eagerly take it up until at least the 1940s. One of its biggest crusaders was American physician and cornflake giant John Harvey Kellogg. Masturbation, Kellogg believed, was a "loathsome ulcer" and society's worst evil. In 1892, he recommended applying blistering doses of carbolic acid to the clitoris as "an excellent means of allaying the abnormal excitement, and preventing the recurrence of the practice in those whose will-power has become so weakened that the patient is unable to exercise entire self-control."†

Some women even requested this surgery. In 1929, in Leipzig, Germany, Marie met a thirty-six-year-old woman who had been "afflicted with compulsive masturbation, repeated up to fifteen times a day, yet was 'totally frigid' with her husband," she wrote in *Female Sexuality*.‡ The woman asked her doctors to be "cured" of her ailment with surgery. After her genital nerves had been severed, both Fallopian tubes and ovaries removed, and her labia minora and clitoris abraded, "she still, however, continued to masturbate just as frequently and compulsively on the scar of the glans clitoris," Marie marveled. "She mas-

---

* More on that in Chapters 6 and 7.

† Oddly, clitoral procedures were also advertised as enhancing female pleasure—specifically, ones to remove the clitoral hood (equivalent to the male foreskin) or free particles from the glans clitoris.

‡ Not surprising, given that "the husband seems to be clumsy in their relations, and carries out coitus without preliminaries," according to the woman's doctor.

turbated exactly where she had always done, with no diminution of 'clitoridal' sensitivity and no gain of vaginal sensitivity."

Marie interpreted this astonishing phenomenon as the stubbornness of the woman's psyche: she must have continued to feel pleasure in her "ghost" clitoris, much as an amputee feels pain in their missing arm.

Importantly, Marie's belief in women's right to feel clitoral pleasure did not extend to all women. Like Freud and others of her day, she held deeply racist ideas about what she considered "civilized" (European) and "primitive" (African and Middle Eastern) peoples. Later in her life, she speculated that traditional genital cutting was meant to repress female sexuality and further "feminize" a woman by removing the vestige of her masculinity. Although she drew parallels between these practices and European clitoridectomies of yore, she concluded that only "primitive" societies still exercised such physical mutilation (she did not seem to be aware that these procedures were still occurring in America).

With anxieties over masturbation at an all-time high, inquiring minds wanted to know: What was "normal" female sexuality? How did it develop? And what did the clitoris have to do with it? Freud, it seemed, had an answer.

When Freud met Martha Bernays, she was peeling an apple at his dining table. It was 1882, long before Marie had entered the picture, and the twenty-one-year-old Bernays had been invited to Freud's family home by his sisters, her friends. Portraits reveal Bernays as a prim young woman, with a thin face and straight dark hair parted down the middle. Her father had recently died, and her eyes were full of melancholy. Freud, in his ample love letters, would describe it as love at first sight.

Freud's first brush with love—which evolved into a protracted courtship lasting more than four years—would make him reconsider his life plan. He was not yet twenty-six, and certainly not yet the father of psychoanalysis. He planned to devote his life to laboratory science. He worked as a research assistant in a physiology lab, publishing papers on the nerve cells of crayfish and the genitals of eels, and harboring dreams of revealing the brain's scientific secrets. But he was still a penniless student, living at home, sporting ill-fitting suits and an unruly beard. What could he possibly offer this woman? Six months after meeting Bernays, he left lab research to become a doctor, a far more stable and lucrative career. He became a clinical assistant at Vienna General Hospital, where he would spend three lonely years trying his hand at surgery, internal medicine, and psychiatry. The path would eventually lead him to psychoanalysis.

It was a woman, in the end, who changed his life course. Reading Freud biographies, one pictures Bernays as a *hausfrau* (housewife) who spent the majority of their marriage caring for their children and keeping their home spotless and welcoming. She was that. But she was also a member of an intellectual class even above Freud's. The daughter of two Orthodox Jews, she was well read and spoke multiple languages. Freud's first gift to her was a copy of *David Copperfield*, and he courted her by sending her poems in Latin, accompanied by a red rose.

After they married in 1886, Bernays would be nearly constantly pregnant for the next nine years, giving birth to six young Freuds. The lengthy love letters devolved into laundry lists. It would be her work in the background—tending the home, raising the children, and entertaining the great minds of psychoanalysis during their Wednesday-evening meetings in her drawing room—that allowed him to follow his passion. Every week, it was only after she had finished serving the men black coffee and cigars that Freud would make his grand entrance. After his death she wrote that it was "a feeble consolation that in the fifty-three years of our marriage there was not a single angry word between us, and that I always tried as much as possible to remove the *misère* of everyday life from his path."

Starting in 1905, Freud began developing a theory of female sexuality that acknowledged society's anxieties but ultimately affirmed that a woman's role was in the home, making babies. His story went like this: A little girl wants to be a man. When she realizes she can't, she rebels, clinging to her phallus and the pleasure it provides her, but ultimately accepts her fate as a woman. "The sexual life of the woman is regularly split up into two phases, the first of which is of a masculine character, whilst only the second is specifically feminine," he wrote in his 1931 essay *Female Sexuality*. "Thus in female development there is a process of transition from one phase to the other, to which there is nothing analogous in males."

The key to this mysterious process was the clitoris. In childhood, this organ was as natural a source of pleasure as a boy's penis; touching it was one of "the child's first, most primal libidinal impulses." But the clitoris was destined to be an infantile organ, inappropriate to the adult woman. It was the relic of a past in which she dreamed of having a penis—a dream she was doomed to give up to achieve the "ultimate normal feminine attitude." "We have long realized that in women the development of sexuality is complicated by the task of renouncing that genital zone which was originally the principal one, namely, the clitoris, in favor of a new zone—the vagina," he wrote.

Freud didn't ignore the science. Trained as a neurologist, he was aware of the growing anatomical understanding of female pleasure. He knew that the clitoris was dense in nerve endings and experienced erection just like the penis, calling it "analogous to the male organ."* He grew up revering Charles Darwin, who had published *Origin of Species* when Freud was three years old. "The great Darwin," as Freud called him, gave him the language to talk about man's will to survive and the universal laws that govern all life. But he eventually felt that he had surpassed even the father of evolution. There were facts of the mind, he believed, that were not subject to biological laws.

Psychoanalysis, after all, was as much about biology as it was about fitting into society's roles. And a woman's role was to accept her lot as a "little creature without a penis"—a castrated male. To function in

---

* In 1908, he wrote that "anatomy has recognized the clitoris within the female pudenda as being an organ that is homologous to the penis . . . this seems to show that there is some truth in the infantile sexual theory that women, like men, possess a penis."

society, she therefore had to reject the clitoris that had given her such joy in childhood and learn to passively receive the penis. What's more, she had to *enjoy* this penetration.

Marie once described the clitoris as a temporary organ, akin to the fetal kidney. It was meant to "eroticize" a young girl, but ultimately step back and let the vagina take over. "Like a deflected stream," she wrote, a woman's libido must "flow through another channel, since it must change its erotogenic zone and pass, in the main, from the infantile clitoris . . . to the definitive adult organ, the vagina."

Freud preferred a different metaphor. In his 1905 book *Three Essays on the Theory of Sexuality*, he compared the clitoris to pine shavings. Its function was to transmit excitement to the "adjacent female parts" just as "pine shavings can be kindled in order to set a log of harder wood on fire." But despite the necessity of this transition, he acknowledged that it took a deep psychological toll on women. "Very often when the little girl represses her previous masculinity a considerable part of her general sexual life is permanently injured," he wrote. "She acknowledges the fact of her castration, and with it, too, the superiority of the male and her own inferiority; but she rebels against these unpleasant facts." That was certainly true of Marie, who was unwilling to give up either half.

Much has been made of the connection between Freud's psychoanalytical theories on women and his relationships with the actual women in his life. What can be said was perhaps best put by American psychiatrist Robert Jay Lifton: "Every great thinker has at least one blind spot. Freud's was women."

Freud readily admitted that women remained a mystery to him. In a 1926 essay (which Marie would later translate into French), he famously called female sexuality a "'dark continent' for psychology."*

---

* This term has racist overtones: Freud borrowed it from Belgian colonizer Henry Morton Stanley, who used it to refer to the continent of Africa as impenetrable, inhospitable, and enigmatic.

Though he claimed to be able to describe these developmental processes, Freud was at a loss to explain their "biological roots." In *Three
Essays*, he wrote that the "erotic life of [men] alone has become accessible to research," while that of women "is still veiled in an impenetrable obscurity." Readers who wished to know more about women, he
wrote near the end of his life, should "interrogate your own experience, or turn to the poets, or else wait until science can give you more
profound and coherent information."

Freud himself, however, consulted a different authority. According
to his biographer, he once confessed to Marie that "the great question . . . which I have not been able to answer, despite my thirty years
of research into the feminine soul, is 'What does a woman want?'"

It was a problem that would vex him for the rest of his life.

———

While Freud approached female sexuality from a theoretical stand-
point, Marie lived it.

She met her first lover during a summer spent hiking in the Swiss
Alps with her father, Prince Roland Bonaparte. Her suitor was her
father's married employee, whom she described in her journal as "the
Corsican secretary, black hair, blue eyes, pointed beard—I was sixteen,
he thirty-eight. I was ugly. He was handsome." The two engaged in an
innocent affair, composed of a single kiss and playing footsie under the
dinner table. One night her suitor requested that Marie bequeath him
a lock of her hair ("To Antoine Léandri, from Marie who loves him
passionately," she wrote along with it). But Léandri wanted more than
her curls. The secretary and his wife conspired to blackmail Marie out
of her fortune, demanding 100,000 gold francs and threatening to take
her youthful love letters to the press.

From that moment, Marie knew that her passion for men was both
her weakness and her strength. "It is not Petite-Maman's jewels in their
box that are my true wealth," she wrote in her diary at the age of

twenty. "It is my heart, it is my mind. And whether or not I am loved or not, I know how to love!"

Marie never fully accepted the fact that she had been born a woman. "Nature and life gave me the brain of a male and the strength and instincts of a male," she wrote to Freud. "I consider myself to have the right to both, and I assure you that it's impossible for me to feel that this is something wrong."

It was only after she married, however, that her sexual exploration began in earnest. In her thirties, alienated from her husband and unfulfilled by motherhood, she began several intense romantic relationships with politicians, physicians, and psychoanalysts. But while she had many lovers, she had little satisfaction. She harbored a deep-seated fear of vaginal penetration, worried endlessly about her inability to orgasm, and fretted that once her lovers had tasted "the mysteries of my flesh" they would lose interest.

In 1924, frustrated by her sexual situation, she found she was in good company. "The number of women afflicted is far greater than men think," she wrote. To get to the root of this mystery, she decided to do field research.

Marie had always longed to study medicine, but her father forbade it as inappropriate for a woman of her stature. Now was her chance. She reached out to doctors she knew in Paris and Vienna, and asked to interview their patients during their gynecological exams. Coming from someone who wasn't as wealthy and well connected as Marie, such a request would have been ridiculous. But because she was who she was, she was able to sit in on hundreds of exams and conduct intimate interviews with women ranging from the age of twenty to sixty-two. In total she queried 243 women on the most intimate parts of their lives—their masturbation habits, whether they achieved a "normal reaction," aka vaginal orgasm, during intercourse—and recorded their genital measurements.

"It then occurred to me to inquire whether something in the genital anatomy of these women might contribute to their defective erotic reactions," she wrote.

Marie noticed a pattern in her data. The size of the clitoris, it seemed, had no effect on whether or not a woman achieved orgasm during intercourse. Instead, the outcome appeared to depend solely on the distance between the base of the clitoris and the urethra. If the clitoris were closer, a woman was more likely to reach orgasm. If the distance were longer, it was unlikely, "as though the gap were too wide to leap."* In the corner of a notebook, she drew a simple diagram of a vagina, looking like a stretched-out lemon, with three different clitorises representing the different possible locations. Marie separated out types of frigidity into those whom she suspected could be helped by psychoanalysis, and those whose problems were purely anatomical—those whose clitorises were the farthest away from their vaginal openings. In 1924, she published her findings in the medical journal *Bruxelles-Médical*, under her male pseudonym, A. E. Narjani.

Then she measured herself. Marie discovered that her own clitoris was too distant to ensure orgasm during intercourse. Here, she believed, was the reason for her frigidity: she was a *téléclitoridienne*, meaning "distant clitoris." For her and other women possessed of such a "stubborn clitoris," she proposed an unconventional solution: "It then occurred to me that where, in certain women, this gap was excessive, and the clitoridal fixation obdurate, a clitoridal-vaginal reconciliation might be effected by surgical means." In other words, why not move your clitoris closer to your vagina, so it could be better stimulated during sex? It was her way of cheating the system, of hacking what she saw as her body's limitations.

Anatomy as destiny? Marie disagreed. Her 1924 paper was a ringing endorsement of a surgery to move the clitoris closer to the vagina, which she helped develop with Halban. She called it the Halban-Narjani oper-

---

* Although it is hardly the only factor, modern studies have reanalyzed Marie's data and found that clitoral-vaginal distance does indeed play a role in women's tendency to orgasm during penetrative sex.

ation. Women's sexual problems, she said, weren't all in their minds. At least in some cases, they were anatomical.

And for her, unlike Freud, anatomy could be changed.

———

A few days after her surgery, Marie awoke to the sound of blackbirds singing in the garden. The weather was perfect, sunny and crisp. On the morning of April 24, her doctor removed her stitches. Things were looking up. "I believe Halban did great work!" she wrote to Freud excitedly. "As for the later results, we'll have to wait!"

She expected to be able to leave Löw Sanatorium by the middle of the week, around the same time Freud departed his own Sanatorium, where he had been recovering from a series of surgeries on his cancerous jaw. By April 28, however, she had not moved. She had gotten a small infection in her surgical wound, which had since healed over. (She oversaw the antivirus treatment herself, she reported proudly.) Still, she was optimistic. "I believe, based on a number of clues, that I'm already able to predict that the operation will be a success," she wrote.

Freud, meanwhile, had still not visited. In a letter, he congratulated Marie on her "heroism" and made it clear he would have no time to see her in recovery. After all, he added, it appeared she didn't need him any longer. His rejection stung her greatly. To be abandoned by the only father figure in her life—the only one who had seen her intellectual and creative potential—was intolerable. After several weeks, Freud relented. But when he finally did visit her, he continued to express his displeasure with her decision. She started to doubt herself. Had she made a horrible mistake?

She soon returned to Saint-Cloud, where her garden was in full bloom. None of it moved her. By the time two weeks had passed, a dreamlike confusion had set in. She was having dreams of God and heaven, though she believed in neither. She felt she had committed a great sin—though she also didn't believe in those either. "There's the sadness in having displeased you, and that which is perhaps even more profound: to have displeased myself," she wrote to Freud. "You awoke me, as we awaken a sleepwalker by grabbing their arm." She was in a neurotic state, she wrote: "There was something in it almost psychotic." Although she was forty-five, she felt like a nervous little girl again.

Much was riding on the results of her surgery. As usual, Marie had mixed work and pleasure. She consoled herself with the thought that soon, she and Halban could collaborate on a paper describing the surgery and observations of other women who had been operated on. If it was a success, other doctors would want to learn it, and the operation would spread. "I'll tell him that only one thing can satisfy me, that we work on what is left to do together," she wrote to Freud. If nothing else, she would have academic fulfillment, her name remembered.

———

On May 18, Marie reported to Freud something remarkable: "*volle Befriedigung*," full satisfaction. "It now happens with great ease, and something interesting: it would be nearly impossible for me to say where the sensation stemmed from, if I didn't know it from the earlier condition," she wrote, alternating excitedly between German and French. It seemed her long quest had ended.

Alas, it was not to be. The experience was either an illusion or a fluke. She would later report that the operation was a failure, and she had never achieved her long-desired conclusion.

Surgery did not give Marie satisfaction. Nor did the Halban-Narjani operation go as planned for the other women who tried it. Marie at first claimed that five women had experienced positive results from the surgery, but later called her original writeup "pre-analytic and erroneous." No women successfully "transferred" their orgasm to their vagina. For most, "the clitoris continued as the dominant erotogenic zone," she wrote. One woman, "thirty-five and a divorcée," was outraged she had allowed the operation in the first place. Before, she had been able to orgasm only while straddling her partner. Now she had an infected cut and no new orgasmic potential. Marie wrote that "this woman's masculinity complex was exceptionally strong," which probably did not help diffuse the situation.

Two years later, however, she was at it again. "Psychoanalysis can at the most bring resignation and I am forty-six years old," she told her journal. "I am thinking of a second operation. Must I give up sex? Work, write, analyze? But absolute chastity frightens me." Noting that "the sensitivity in the place from which the clitoris had been moved persisted," she resolved to go under the knife once more. In 1930, Halban came to Paris to operate on her again. For some time, Marie had also been experiencing painful swelling of her Fallopian tubes, known as salpingitis, which left her bedridden for weeks at a time. Halban ended up not only moving her clitoris but also removing her uterus and ovaries, sending her on an accelerated track to menopause.

In light of her experience, she continued to build on her theory of female sexuality. Yet to Freud's chagrin, she refused to repudiate her clitoris. Instead, she became solely focused on her phallus and its potential for pleasure. "Woman's share in sexual pleasure seems to be derived from whatever virility the female organism contains," she wrote. The little girl, she wrote, must have at the outset some "residue of virility" in order to "eroticize her"—but not too much, lest she become overly fixated on her clitoris and unable to transfer her pleasure to her vagina.

In November 1934, Marie traveled to Buckingham Palace. She was fifty-two, the corners of her eyes beginning to crinkle, her coiffed hair streaked with gray. Officially, she was here to attend the wedding of her niece Marina, daughter of Prince Nicholas of Greece, to the Duke of Kent. But during the wedding, she snuck off to the London offices of the British Psycho-Analytical Society to give a lecture on women's sexuality. Before a mostly male audience, she advocated unabashedly for the "harmonious collaboration" of clitoris and vagina. She spoke of the female fear of penetration, and of the pleasure women receive not from some "particular zone" but from their body as a whole.

Marie led a double life. To her royal family, she was Auntie Marie, a doting old lady and "splendid old thing" in the words of one psychoanalyst friend. To her professional colleagues she was a respected, if eccentric, expert on female sexuality. Given her lifelong quest to find pleasure in penetration, Marie is often remembered in terms of longing and unfulfillment; her fixation on achieving orgasm even led the sculptor Constantin Brâncuşi to create a shining bronze phallus in her honor known as "Princess X." But this is not quite right. Marie sought not to forgo her clitoris and dedicate herself to passivity, but to bring two worlds together. "Humanity," she wrote in *Female Sexuality*, "does sometimes know how to reach happy compromises."

Yet in her own life, she had not achieved that unity. While she preached erotic harmony, her sense of a divided self persisted. This division can be read in her dual identities, male and female; it can be read in her life path, aiming to be a doctor but ending up a bearer of heirs; and it can be read on her physical body, scarred by a surgery she deemed necessary to meet the gendered norms of her time. Her path was a deflected stream, seeking intellectual fulfillment but always thwarted by society's demands.

For her, surgery could not bridge the divide between her two selves. But neither, it seemed, could psychoanalysis.

⌐⌐⌐

Scholars have lamented that Charles Darwin, the theorizer of natural selection, had the misfortune to never cross paths with monk-turned-botanist Gregor Mendel and encounter his work on the genetics of pea plants. While Darwin was alive, Mendel provided a mechanism for his theory of natural selection, rooting it in the transference of genetic material through discrete units now known as genes. Meanwhile, Darwin spent his life guessing at what lay behind his great theory of inheritance, but never quite getting there. Similarly, for all her reading, Marie never came across the work of, say, anatomist Georg Kobelt. If she had, she would have understood her mistake in an instant.

What limited Marie wasn't her lack of understanding of psychoanalysis. It was her—and her doctor's, and the rest of the world's—lack of understanding of female anatomy. Marie thought of her clitoris as a diminutive penis. But what she considered her jewel, it turns out, was merely the tip of the iceberg—the glans clitoris, equivalent to the head of the penis. Just as the penis has a root of nerves that adds one-third to its length and extends deep into the pelvis, the clitoris, too, sits atop an unseen empire, a palace of nerves and blood vessels all connected beneath the surface.

Freud was wrong: The clitoris and vagina were not two separate entities, destined to exist in constant conflict. They were interwoven in their nerves and fates, intimately intertwined. They were part and parcel of the same thing. The clitoris is an underwater volcano, a pyramid mostly buried in sand, a jellyfish-spider reaching its tingling tendrils into every crevice of the female pelvis—an edifice of powerful sensory material that Bonaparte's surgery never touched.

Erotic harmony, it turned out, already existed inside of her. She just hadn't known how to look.

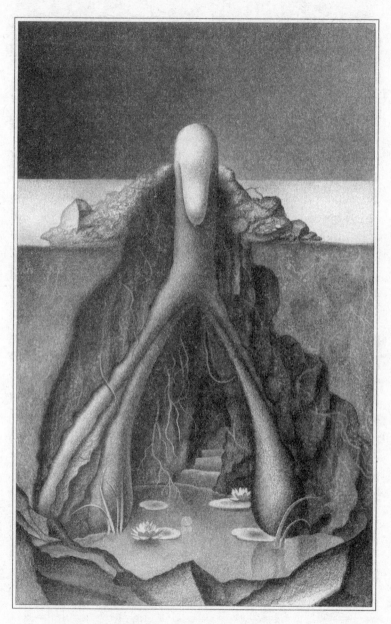

We only see what we choose to and disregard the rest.

—HELEN O'CONNELL, 2004

CHAPTER 2

Wholeness

(INTERNAL CLITORIS)

" It's really something, isn't it?"

Dr. Helen O'Connell's voice is tinged with awe—reverence, even. The fifty-nine-year-old Australian urologist is gesturing to what at first looks like a satellite image of the surface of Mars, full of pinkish-red rivulets and craters. Actually, it's terrain less known than the surface of the Red Planet: a cross-section of the body of the clitoris, showing the microscopic structures within the tissues. She points to the heart-shaped structure in the middle, split into two hemispheres, encased in a thick red lining. It's porous, full of tiny caverns. "See how vascular it is?" she says. "This is all erectile or spongy tissue." When a woman becomes aroused, these spaces fill with blood, just like the columns of erectile tissue in the penis.

Within the clitoral body are whitish, Impressionist swirls that push up against each other, like the spiraling gusts of wind in Van Gogh's *Starry Night*. These dense nerve endings are what make the clitoris so exquisitely sensitive to the touch. They are tightly packed in the glans clitoris and the erectile bodies, where they respond to light pressure and vibration. On either side of the body are two red-ringed craters: the clitoral arteries, which pump blood to this sensation-rich part of the pelvis. Bundled alongside them are what O'Connell stresses are sizeable nerves: even in an infant their diameter is 2 millimeters, visible to the naked eye.

"Look how pink the nerves are," she says. "That's really amazing, isn't it? So close to the tip but there's nerve trunks . . . Instead of fibers, which you wouldn't see, that's a huge intact nerve."

This is page 133 of O'Connell's doctoral thesis, "Review of the Anatomy of the Clitoris," bound in red leather. Published in 2004, it is perhaps the most thorough exploration of the anatomy and history of the human clitoris ever written. She's dusted it off in anticipation of our meeting at Western Hospital at Footscray, a sprawling medical complex five miles west of Melbourne, in late February 2020. (This was days before Covid-19 hit Australia, and she was working at the hospital as usual.) O'Connell is head of urology here; today, instead of surgical scrubs, she's wearing a smart black pinstripe suit, her blue eyes alert behind her signature maroon cat-eye glasses. As she flips enthusiastically through the pages, these alien images begin to feel like familiar landscapes, maps of a place you've come to love.

When it comes to understanding the anatomy of the clitoris, all roads lead to O'Connell. Throughout her career, she had seen this body part systematically shafted—from Vesalius to Freud, up until her modern medical classes. Typically, her textbooks and teachers cast it as the smaller, inferior cousin to the penis. But she never believed it. It's true, she admits, that the clitoris is harder for researchers to examine, given that most of its bulk is shielded by fat and hidden under the arch of the pelvis. But she always had a hunch there was another reason: they just weren't looking for it.

"The original anatomists weren't interested in the clitoris. The penis was much more interesting," she once told the BBC. "It was bigger and you didn't have to wear your spectacles to see it."

The first three years of O'Connell's medical training were spent memorizing an endless list of Greek and Latin names for body parts. That was bad enough. But then there was "this hideous book." The

1985 version of *Last's Anatomy*, the textbook assigned for her surgical exams, "was written in this sort of low-key style that was a bit offensive to a young woman," she recalls. The clitoris barely got a cameo; a cross-section of the female pelvis omitted it entirely. Meanwhile, four pages were devoted to the penis. The text referred to aspects of the female genitals as a "failure" of male genital formation. She underlined that word in blue pen, along with phrases like "lacking the rigid support," "poorly developed," and "no glands," in reference to the vagina.

"No mention of changes in its structure in sexual activity! Or erectile mechanism," she scrawled in the margin.

By this time O'Connell was in her mid-twenties, married. "I knew from my own body that there was something there, right? *Last's* had nothing. That's wrong. And then what was there, what is the anatomy, is really an open question."

Later, while working at a women's health center in Melbourne, she started rubbing elbows with researchers outside her field, biologists and psychologists and social scientists. One turned her on to a very different type of anatomical resource, a women's health book called *A New View of a Woman's Body*, which illustrated the clitoris as a long, winged body. The authors, a group of American women who had formed a feminist health group, lacked access to cadavers for dissection. So instead, they'd stripped from the waist down and examined each other's vulvas.* They compared what they found to images in standard anatomical texts, noting the wide variation among women. "Amazing methodology," says O'Connell.

She filed it away in the back of her mind. She had yet to complete her urology training, which would make her the first female urologist in Australia. "But I definitely made some sort of internal resolve to look at it further when I got the chance," she says. "In the back of my head was this intention."

---

* One participant even masturbated to orgasm in front of the illustrator, so that she could see the clitoris and its surrounding structures in each stage of arousal.

Once she started her career in the male-dominated field of urology, the problem became clearer. For men undergoing prostate surgery, surgeons were careful to avoid any nerves or blood vessels that might, if cut, interfere with their sexual lives. For women, it was a minefield. No one knew exactly what nerves were down there, or how they connected; no one had considered the delicate architecture of their bodies in the same way. When it came to urethral surgeries, hysterectomies, forceps deliveries, and procedures to remove mesh slings from the pelvis, the clitoris was in the firing line. "For a surgeon," she said in 1998, "that's unacceptable."

By the end of medical school, O'Connell was determined to find out what was really going on down there. The answer, she suspected, would lie somewhere between the texts penned by feminists and the ones penned by male anatomists. Like Vesalius, she decided to work it out not from the books and old masters but by going back to the basics: dissecting bodies. Lots of bodies. "What was the truth?" she says. "You have to look at it."

In 1998, O'Connell and her colleague, microdissector Robert Plenter, were hunched over the pelvis of a cadaver, sawed through at the mid-thigh and lumbar spine—"a bit like a pair of shorts," as another dissecting colleague put it—at the University of Melbourne. She was dressed head to toe in blue scrubs and plastic goggles, her blond hair pulled back. Her hands were encased in two sets of latex gloves in a futile effort to keep the acrid smell of formaldehyde from seeping in. Above her whirred a metal fume hood, which sucked up the toxic fumes from the fixative used to prevent tissues from decaying too quickly.

Plenter, whose expertise was in tiny surgery, specifically heart transplants on mice, was showing her a clitoris. But not just any clitoris. It belonged to a woman in her thirties, tall and strong, who had apparently died of a facial tumor. The other bodies they had been working with were elderly, meaning their clitorises had shrunken somewhat with age. (Most were also fixed in formaldehyde, which shrinks tissues down further.) More than any of their past specimens, this one represented what young, healthy clitoris might look like. It was full of erectile tissue and branching nerves. "As much as cadaveric work can be beautiful . . . she was so perfect," O'Connell says. One colleague recalls being shocked to find that there appeared to be as much erectile tissue in the clitoris as there was in the penis, just distributed differently.*

This unfortunate woman was exactly what they had been waiting for: their crown jewel, confirmation of an emerging understanding they were coming to based on their dissections of older bodies.

Starting in 1996, O'Connell and Plenter worked in this small dissecting room on the fifth floor of the university medical building, searching for anatomical truth. Systematic and methodical, the pair removed layer by minuscule layer, like archaeologists discarding dust and dirt to get at the shape that lay beneath. They had to go slowly, because they were treading on what was essentially new territory to modern science. It could take hours to dissect millimeters of flesh, and months to dissect a full pelvis. When they got to the fine dissection, they used loupes—magnifying lenses mounted on glasses used by dentists and surgeons—and took an almost constant stream of photographs. "I'm known for my patience, fortunately," Plenter says.

What emerged was no single organ but a cluster of erectile tissues that wrapped around the vaginal canal and urethra. The shape was of a phallus pointing down, with two arms swooping up to nine centimeters (3.5 inches) back against the pubic bone, and a pair of teardrop-shaped bulbs hugging the vaginal canal and the urethra. This was not a struc-

---

* O'Connell, for her part, bemoans the obsession society seems to have with comparing female genitalia to male genitalia. To her, it's part of the ancient trap of seeing the male body as primary, and the female as a deviation. Most textbooks inevitably introduce the penis first; the clitoris is described in terms of how it either parallels or diverges from that norm. She wishes people could just take the female body on its own terms, and appreciate the clitoris as its own, marvelously shaped, impressively sized structure.

ture that could be imagined on a flat plane, in two dimensions, the way it was in textbooks, if it was represented at all. You had to see the way it wrapped around the vagina, and flared out into the surrounding tissue of the pelvis. You had to be able to picture the visible part—the nub that Marie Bonaparte had thought of as the full clitoris—as the very tip of a pyramid of erectile flesh, the broad base obscured by sand.

As they dissected more bodies, O'Connell grew more confident in the shape. But to be sure, she cross-checked her findings with a set of magnetic resonance imaging (MRI) images that visualized this area in ten healthy, younger women. The pictures revealed an astonishingly high blood flow in this area, making the clitoris glow bright white. In 2005, she published a paper that wove together literature review, microdissection, and MRI imaging to lay out the full anatomy of the clitoris from every angle. The vagina itself, it appeared, was not erectile or particularly sensory in nature; rather, it was the clitoris's close relationship to the vagina that gave it its role in sexual pleasure and arousal.

This was a paradigm shift. At the time, many medical textbooks still referred to the glans as if it were the full organ. In reality, O'Connell wrote, "the glans is a small, button-like extension of the body of the clitoris," similar to the head of the penis. The full clitoris was ten times the size most people thought it was, and shaped like a mountain of erectile tissue rather than what its Greek name suggested, "little hill."* More than that, it was intimately connected to all the organs around it.

O'Connell's work showed definitively that Freud's dichotomy between the vagina and the clitoris was false. He and other psychoanalysts had instructed women to give up their "infantile" clitoral pleasure, and embrace their vaginas. But any sensation you had from vaginal penetration, her findings suggested, came from the stimulation of the different parts of the clitoris through the vaginal walls. It was just as American feminists and, in a way, Marie Bonaparte had said: Almost every type of pleasure could be

---

* The same Greek word is associated with the meaning "to rub," leading some scholars to suggest an ancient play on words—but no one knows for sure.

explained by the clitoris. There was no clitoris versus vagina; they were one and the same. O'Connell coined the term "clitoral complex" to refer not just to the glans but to the interconnected network of tissues that shared a blood and nerve supply, and responded as a unit to arousal.

Her discovery struck a nerve—pun intended. Media deemed her a pioneering anatomist and "clitoris guru" who was changing the way we thought about women's bodies. "Penis envy may be a thing of the past," wrote the magazine *New Scientist*, after the publication of her first paper. "The clitoris, it turns out, is no 'little hill' . . . It extends deep into the body, with a total size at least twice as large as most anatomy texts show." Finally, her work had restored "the clitoris to its rightful place on the body's map." (Her old anatomy lecturer, Dr. Norman Eizenberg, has incorporated her findings into his virtual learning program, Anatomedia. Her "perfect" clitoris lives on there as the main specimen, "in all its glory," he says. There's even a plastinated clitoris in the university's anatomical museum, with the glans labeled as "external tip of clitoris" and the rest labeled "clitoris.")

She may not have been Vesalius. But she had just made a discovery that would have laid the father of anatomy's understanding of the female body to rest.

O'Connell never intended to be the second coming of Vesalius. She was just trying to address some of the most glaring deficiencies she had encountered in medicine. "There's a project that needs to be done," she says. "I don't consider myself to be an anatomist. I'm a doc, urologist, surgeon, who had the opportunity to look objectively at something and contribute."

O'Connell thinks of herself as using modern science to confirm Kobelt's landscapes, surfacing them back into the limelight where they belong. But Kobelt hadn't gone the full distance. Although he had illustrated the clitoral bulbs, he called them "bulbs of the vestibule"—the "vestibule" being sort of a waiting lobby to the vagina—and referred to them collectively as the "passive female sex organ." Moreover, he overlooked the relationship of the clitoris to the urethra and other surrounding structures. O'Connell could see how all these parts worked together, which helped explain why urethral surgeries risked damaging clitoral nerves. She went past Kobelt's understanding, unifying all of the parts of the clitoris into one integrated superstructure.

Like the clitoris itself, her findings radiated out into the cultural milieu. They found their way into the minds of OB/GYNs, artists, jewelry-makers, and academics, who drew on her research to reimagine the shape of female pleasure. Her data has been used to create three-dimensional models of the clitoris that are used by some of the people most invested in this anatomy: sex educators and therapists, gender-affirmation surgeons who shape vaginas and clitorises out of penises, and doctors who perform surgeries reconstructing the anatomy of women who have undergone genital cutting. They use these models to understand the anatomy in three dimensions—and to help their patients better appreciate their own bodies.

───

Aminata Soumare was seventeen the first time she heard the word. She was sitting in a biology class in her second year in *lycée,* the public secondary school in France, in early 2017. Her teacher had separated the class by gender, and now she and fifteen other girls were about to learn about the female genitalia. The teacher put an up-close diagram of the external female genitals on the screen and began explaining the different parts: labia majora (outer lips), labia minora (inner lips), vaginal opening, urethra. Above the vaginal opening, where the two smaller lips came together, she pointed out a shiny pinkish nub with a fold of skin over it. That's when she said the word Aminata had never heard of before: clitoris.

The what?

"I looked and thought, Why is there all of that on the photo? I don't have any of that!" Aminata recalls. She felt lost. "It's a bit weird to see a photo that says, 'This is what everyone should look like' and to realize that's not what you look like." Today, Aminata is a round-faced twenty-one-year-old with full lips and black-eyeliner-rimmed eyes. The day we first spoke, in November 2019, she was wearing a small silver nose stud, her dark, curly hair pulled back in a ponytail. She spoke like a teenager, using the French equivalent of "like" and "you know," each sentence tumbling excitedly into the next.

That day in 2017, questions flooded her mind, but she didn't dare ask her teacher, not with everyone watching. Instead, as soon as she got home, to the apartment in the suburb of Saint-Denis that she shared with her mother and three siblings, she went to her bedroom and typed "clitoris" into the Google Images search bar. Row upon row of pictures of female genitals popped up. Then she sat on a chair in front of her full-length mirror and looked between her own legs. In the place where those women had a pinkish nub and a fold of skin, she had nothing. It was more of a dip, a valley—"*un creux*," she called it. A hollow.

Talking to her mother wasn't an option. They didn't have the kind of relationship where they could talk about things like sex. Instead, Aminata approached her younger sister, and asked her what she had "down there." It was the same with her. They realized the same thing must have been done to them.

In that same biology lesson, the teacher had shown a video about an African woman who had undergone a drastic form of female genital cutting.* In her case, her labia were sewn together leaving only a tiny

---

* The World Health Organization refers to these practices collectively as "female genital mutilation," defined as "all procedures involving partial or total removal of the external female genitalia or other injury to the female genital organs for non-medical reasons." Yet scholars have pointed out that this term is inappropriate for several reasons. First, it's inconsistent: By that definition, it should also encompass consensual surgeries like labiaplasties (which are performed on adolescent girls in the United States and elsewhere), vaginoplasties, and genital piercings. Second, it's imprecise: The term covers a wide range of practices, ranging from a "nick" that does not involve removing healthy tissue, to procedures as severe as infibulation. Finally, some women who have undergone cutting find the term "mutilation" stigmatizing and harmful. Therefore, I've chosen to use the less culturally loaded—and, some argue, less Western-centric—"female genital cutting."

opening for urination and menstruation, a practice known as infibulation. The film had made her shudder—thank God that hadn't happened to her, she'd thought.

But now, she thought, maybe something similar *had* happened to her.

———

Aminata has lived in France since she was twelve years old. But she was born in Bamako, the capital of Mali. After her biology class, she would learn that Mali is one of the few countries in Africa where female genital cutting is still legal and widespread. According to a 2019 UNICEF report, around 8 in 10 girls in Mali have undergone genital cutting. In Bamako, the capital in the southwest where Aminata is from, the rate is even higher. Girls are usually cut early in life, before the age of five. Aminata couldn't wrap her mind around it. On her vacation back to Mali later that year, she decided to ask her grandmother the question that had been on her mind ever since she found out about the practice: why do they do this to women?

"Well, it's a tradition," she remembers her grandmother saying. "They've done it to us ever since we were little, and everyone says that when we do excision the woman doesn't turn into a whore."

"But what's the point?" Aminata pushed her. In Mali, girls can legally marry with parental consent at age fifteen if a civil judge approves, and marriage as young as ten is not uncommon. If women get married so early, why would there be a worry she would become a loose woman?

"Listen," her grandmother replied, "it's not us who decides, and it's always been done."

One answer may lie in the French language itself. In most other parts of the world, people use the term "female genital mutilation," or cutting. But in France, the common term is *mutilation sexuelle*, or sexual mutilation. (The other umbrella term is "excision.") This term makes a distinction between the clitoris—which, as the only organ devoted solely to pleasure, is not usually considered part of the reproductive system—and the reproductive organs. "This 'operation' doesn't touch the reproductive organs, the genitals," says Sokhina Fall Ba, former co-president of the End FGM Network, who has worked with excised women in France, Mali, Burkina Faso, and elsewhere. "It doesn't stop a woman from having a baby. But it does attack sexuality."

The term "sexual mutilation" gets across the fact that, although genital cutting practices span many cultures, they are often united by a shared theme. It's the same reason similar practices were used in Europe and the United States in the time of Marie Bonaparte, when figures from Isaac Baker Brown to John Harvey Kellogg touted clitoral amputation as a cure for the scourge of masturbation. The idea was that, by removing the part of a woman that gives her pleasure, it would diminish her desire while rendering her still able to execute her reproductive function. "Our ancestors were no scientists," wrote journalist Esther Ogunmodede in *Drum*, a Nigerian magazine in 1977, "but they knew where and what was the seat of sexual pleasure in a woman—so they chopped it off before they could discover it."

When Aminata returned home from Mali, she finally got up the courage to ask her mother if she knew what excision was. "Yes, I know what it is, because I'm excised too," her mother said. Like Aminata and her sister, she was cut soon after birth, too young to remember.* That wasn't enough for Aminata. Soon after, she made an appointment with her gynecologist, who examined her and confirmed that yes, the tip

---

* Other women I talked to remember everything. A scythe-shaped blade. The screams, the smells. The blood. "I remember it as if it happened today," I was told by Corinne, a twenty-five-year-old woman from the Ivory Coast. "You pray that it's over quickly. You have no choice." When Corinne was eleven, she remembers being held down, a woman grasping each of her limbs and forcing her legs apart, and another woman, the cutter, crouched between her legs. "What can you say? Even if it hurts, it's an obligatory rite of passage."

of her clitoris had been removed. Then she told her about a place she could go, an organization called La Maison de Femmes, the house of women, that specialized in helping women victims of violence. There they could give her more information. Maybe, her gynecologist said, they could even undo what happened to her.

———

La Maison de Femmes is a pop of color in an otherwise dreary landscape. The main building is shaped like a large house, with a sloping roof. One wall is a vibrant magenta, another lemon yellow, another apple green. Surrounded by a field of grass and a high chain-link fence, La Maison serves all victims of gender-based violence, including forced marriage, rape, and incest. It is located in Saint-Denis, part of the Seine-Saint-Denis *département*, the poorest county in metropolitan France. Many who come here are recent immigrants, fleeing violence in their home countries, facing citizenship difficulties and lacking the legal right to work. Hundreds of women of French nationality come each year as well. Domestic abuse, which spiked during France's many lockdowns during the 2020 pandemic, has long been a silent problem in the country.

Aminata had walked past the building many times but had never been inside. Now, on the day of her first appointment, in January 2019, she was nervous. She had brought along a friend from school who had emigrated from Senegal, and who also thought she had been excised. They rang the buzzer. The lobby was open, the entrance framed by exposed wooden beams that sloped upward toward the sky, and sunlight filtered in through skylights and floor-to-ceiling windows. There were walls of fluorescent pink, and a multicolored sculpture of a dancing Venus of Willendorf. In the corner was an area for young children, with books and plastic furniture. A yellow strip on the floor read PLEASE RESPECT THE ZONE OF CONFIDENTIALITY in French.

After they sat down in the front lobby, a nurse came over and

.gently told them they were in the wrong place. She directed them to the back waiting room, which is for women who have undergone genital cutting, along with those seeking abortion counseling or family planning. It was plainer there, the walls decorated with a few abstract paintings of red and green confetti, and a big red couch looking out onto an empty field. It was quieter too. The voices from the front desk were muffled. The only sounds were women talking in low voices in several different languages, and sometimes the soft babbling of a baby in a stroller. Every so often a staff member in a white lab coat came out of an office to call out a woman's name: Criztena. Corinne. Myriam.

Aminata was called in first. Dr. Ghada Hatem, a small, jovial woman with light-blue eyes and a lion's mane of thick bronze hair, ushered her through the open door of her consultation office, which featured a colorful portrait of Frida Kahlo. She asked Aminata to sit in a lime-green chair with stirrups with her legs apart, so she could examine her. Hatem leaned in closely.

It didn't take long for her to finish the examination. Yes, she told Aminata, she had been excised. But why, exactly, had she come here?

Aminata explained how she had found out about what had been done to her during her biology class. She didn't have any medical problems, besides frequent urinary-tract infections. But her dip bothered her. Now that she knew what had been done to her, it began to feel like a constant irritant. "That was my daily nightmare, when I woke up that was the first thing I thought about," she said. "It bothered me, it made me self-conscious even, because it made me think that I'm not like everyone else, something's missing."

Next, Hatem asked Aminata if she had had sex. She had; she had a boyfriend of about a year. Did the excision bother her boyfriend?

Aminata told her that her boyfriend didn't even know she was excised until she told him. Aminata was the one who was bothered by her inability to feel anything during intercourse or while touching herself. "You feel like you're like a zombie," she said. "There's no point in going out with someone. Of course, it depends on the person, but for me that really bothered me."

Hatem listened. She showed Aminata photos of women who had been excised and those who hadn't. Then she reached into her desk drawer and pulled out a three-dimensional model of an anatomically correct clitoris, fluorescent pink, its winged body embedded in a rubbery silicone vulva. This was an educational model made with the data O'Connell had gathered, twenty years earlier. Hatem uses it to explain to her patients the true shape of the clitoris. Today, she used it to show Aminata what part had been cut, and what remained.

Aminata was shocked by how large the model was. The nub she had seen in class—the part the cutter had removed—was only the tip of the iceberg. The real thing was this wraparound 3-D shape, an underground kingdom. It had wings and bulbs. It looked like a penguin, or a spaceship.

What most people don't realize, Hatem explained, is that the clitoris is far more resilient than it's been given credit for. About 90 percent of the organ, and most of its nerve endings, are under the surface. That meant Aminata still had most of it within her—it was just buried. Next Hatem told her how, in a simple operation, she could feel around for the remaining clitoris under the skin, free it from the pubic bone, and bring it back up to the surface to expose the nerve endings. (Aminata's friend would turn out to have had only her labia cut, meaning reconstruction would not be useful to her.)

Aminata felt herself grow light. Once she realized there was another option, she felt like she couldn't keep going as she had before. Even after she had gone home and looked up videos of the operation on YouTube and seen how bloody it was, she wasn't deterred. *It will be a proper operation*, she told herself. *Not like what was done to me when I was little.* She knew she wasn't going to be perfect, or like she was before. But as she put it, "I'm going to be a bit more like everyone else, even if I know that I won't be exactly like everyone else."

I want to feel whole. I want to feel normal. I want to have my identity back.

These are the words that surgeons like Hatem hear over and over again. Listening to them, it's clear that this small bit of visible flesh often represents something far larger. Reconstruction is about wresting back not just your body, but your autonomy, your identity. Her patients want to recover some part of themselves they feel they have lost.

In 2013, Hatem was director of the maternity ward at the adjacent Centre Hospitalier Delafontaine, a gray-and-brown building you can see from the windows of the House for Women. That's where she began to see it again and again: scarred-over clitorises, women without visible clitorises, women with cuts and scars to their labia. Some didn't know they had been cut until she told them. She had seen this before in other hospitals, but it had always been *ponctuel*—from time to time. Now it was every week. Hatem herself had emigrated from Lebanon in 1977, to flee her country's civil war and study medicine in France. "Having known war made me intolerant of all forms of violence," she told the French newspaper *Le Monde* in 2021.

She did an audit of the patient data and found that out of the 4,700 women who delivered at her ward per year, between 14 and 16 percent had been cut. Something, she realized, needed to be done.

She began to imagine a place these women could go. It would be a welcoming place, bright and cheerful, compared to the bleak sterility of the hospital halls. A place with not just doctors but sexologists, therapists, nurses, midwives. A place where care was fluid, and women could come for all their needs—emotional, sexual, medical—so they didn't slip through the cracks of the French healthcare system. A *parcours*, she called it: a pathway, or process.

Importantly, all of its services would be free. La Maison is a charity attached to the hospital, meaning those who enter its doors pay nothing. That includes procedures like clitoral reconstruction, which promises to repair the clitorises of women who have suffered genital cutting. In 2004, the French parliament decided to cover reconstruction under the national health service for women who experience pain related to their cutting; it has since expanded to cover women wishing to improve their physical appearance or sexual sensation. That the surgery is free is important. By the time Aminata decided to get the operation, she had finished high school, and was living with her sister and mother while training to become an assistant nurse. "I wouldn't have been able to do it otherwise," she said.

Today, about a third of the women who come to MDF come because they have been cut. Their injuries range from a "nick" to the labia to the removal of the glans clitoris to the sewing up of the vaginal lips, which can cause severe problems with giving birth. And since the government's decision to cover the surgery, the number has been steadily increasing. Hatem performs at least one hundred clitoral-restoration surgeries per year. In 2021, after more than a year of pandemic-induced delays, she was able to expand her facilities to accommodate the rising tide of women seeking her services.

In her consultations, Hatem cautions women not to expect too much from reparative surgery. The operation is not an undoing, and no procedure can them give back exactly what they lost. She doesn't promise sexual pleasure; she doesn't promise orgasm. "It's better than nothing, but it's not perfect," she says. Dr. Pierre Foldès, another French surgeon who is considered the original developer of the surgical technique to repair the clitoris, echoes this caution. Foldès believes the outcome of the surgery should be judged not by the anatomical result but by how a patient feels about her body afterward.

In the 1980s, Foldès was a young humanitarian doctor working in Burkina Faso. He was mainly repairing rectovaginal fistulas, abnormal openings between the walls of the rectum and the vagina that can occur after a long or obstructed childbirth. That's when he started to encountered women who had been cut. "The first request was pain," he says. The problem was that Foldès, who, like O'Connell, had trained as a urologist, knew little about the clitoris. Actually, no one did. So he went back to France and began dissecting cadavers, using ultrasound to scan living women's pelvises, and scouring historical anatomical textbooks. He, too, came across the striking illustrations of Kobelt.

Eventually, he realized that these nerves had merely been buried. They could be pulled up and enlivened. He came up with a simple technique to remove scar tissue and pull the untouched part of the clitoris back up where it belonged—the same technique he would later teach to doctors like Hatem. Later, he would also learn to rebuild the labia, which were often damaged, and create a new clitoral hood from a fold of skin. When he returned to Burkina Faso, he could tell these women: yes, something could be done.*

Today, most surgeons who perform this surgery initially learned it from Foldès, who estimates that he has operated on at least 6,000 women. But surgery, he knows, is not a panacea. Sexuality is about far more than clitoral sensation; it's about how we feel in our bodies, how we see ourselves, and how we communicate with sexual partners. For many of his patients, the most meaningful part of the process is the holistic care they go through along the way, speaking with sexologists, therapists, and other experts—the *parcours*. During that journey, some find that their lack of pleasure is not directly connected to the cutting, says Hatem. "Sometimes it is," she says, "and sometimes it's only because their parents never explained anything about sexuality or about mutilation, and they never touched themselves, and they know nothing and they feel shame."

The mark of a successful surgery, Foldès believes, is simple: "women

---

* In our conversation, Dr. Foldès often referred to female sexuality as being either "clitoridian" or "vaginal." When asked what he meant, he clarified that by "vaginal" women, he meant those who feel more pleasure from the arms and bulbs of the clitoris, rather than the glans. But his words gave me pause, because they were the same ones Freud popularized more than one hundred years ago to categorize women as either "infantile" or "mature." Even today, Freud's ghost looms over the clitoris.

feel the fact of being normal." But what is "normal" when it comes to female sexuality? The question isn't a new one. Even in Marie Bonaparte's time, researchers were starting to ask it. For decades, Victorian morals and Freud's reductive framework of vagina versus clitoris had stymied progress. But starting in the 1930s, American sexologists and gynecologists began challenging Freud's unscientific theories, and turning to the clitoris to better understand the anatomy of female sexual pleasure. Nerve by nerve, they began to restore the clitoris's lost reputation.

———

One of the first researchers to turn clitward was Dr. Robert Latou Dickinson. A mild-mannered marriage therapist and gynecologist with a penchant for statistics, Dickinson probed "the everlasting gonad urge," taking in-depth case histories of thousands of patients at his gynecological practice in Brooklyn. Appalled by how little research had been done into "normal sex life," he made it his goal to start the fledgling field of sex science off with a foundation of hard data.[*]

His immediate goal was to help married couples—mainly white, heterosexual, middle- to upper-class ones—achieve sexual harmony.

[*] To be clear, others were spreading the clitoral gospel long before Dickinson—their work just wasn't necessarily seen as science. In 1918, a thirty-eight-year-old British paleo-botanist named Marie Stopes published a frank, woman-centered sex manual after her own marriage was annulled due to failure to consummate. In her wildly best-selling pamphlet *Married Love*, she not only laid out an early version of the principle of affirmative consent ("a man does not woo and win a woman once and for all when he marries her: he must woo her before every separate act of coitus") but also spelled out the importance of the clitoris in mutual sexual pleasure. Like the penis, she wrote, the clitoris "is extremely sensitive to touch-sensations. This little crest, which lies anteriorly between the inner lips round the vagina, enlarges when the woman is really tumescent, and by the stimulation of movement it is intensely roused and transmits this stimulus to every nerve in her body." Unfortunately, like her American colleague and fellow contraception crusader, Margaret Sanger, Stopes was also an avowed eugenicist.

That meant trying to solve a range of bedroom problems, from so-called frigidity to premature ejaculation. But he was particularly interested in the female side of things. The more women he looked at, the more he was shocked by how rarely women seemed to be satisfied during conjugal sex. Only 2 percent, he concluded in his 1931 report *A Thousand Marriages,* experienced regular orgasm with their spouse. (Most of these couples, he noted, were primarily utilizing the missionary position, which might have something to do with it.) This contrasted strikingly with his finding that two-thirds of women masturbated, presumably with satisfaction.

To find out what was normal when it came to sexual anatomy and behavior, he set out to acquire "exact genital measurements." This meant doing vaginal examinations and taking measurements of thousands of patients.* A breathtakingly talented illustrator, Dickinson sketched his patients' intimate anatomy with pencil and tracing paper, meticulously shading for texture, trying to capture what happened to the genitalia in every position and state. These illustrations fill half of his 1933 *Atlas of Human Sex Anatomy,* dedicated to "the science—and art—of sex life." The book overflows with observations and drawings that reveal the astonishing diversity of human sexual anatomy, from labia minora to glans clitoris. His artistic talents once earned him the nickname of the "Rodin of Obstetrics."

His research convinced him that the clitoris was the underappreciated center of female pleasure. "The female organ is minute compared with the male organ," he wrote. "But the size of its nerves and the number of nerve endings in the glans of the clitoris compare strikingly with the same provision for the male . . . the glans of the clitoris is demonstrably richer in nerves than the male glans." (Note: While one modern study has suggested the clitoris may be more densely innervated than

---

* Not all of these observations were taken ethically. Dickinson reportedly supplemented his written notes during examinations by using a secret camera, hidden in a flowerpot, which he could discreetly activate by pressing a switch.

the penis, no one has actually compared the two.) In detailed images, he showed how the clitoris could experience erection, becoming a "half pea-sized protrusion of firm rotundity" that could "yield to the touch a regular throb."

Though he championed the clitoris, Dickinson was no radical. A devout Episcopalian and devoted father of two daughters, Dickinson was particularly concerned with what comprised a "normal" sex life. "We in medicine . . . are called upon to do our part to persuade morals to wed normals," he wrote. Through most of his career he held the unfounded belief that "deviant" sexual acts like masturbation, promiscuity, and lesbianism could enlarge the clitoris. In a 1940s sculpture entitled *Forms of Adult Vulva Drawn Open*, Dickinson labeled a vulva with loose, wrinkled lips as a "masturbator" and another with an enlarged clitoral hood as "homosexual."

In the 1940s, he and sculptor Abram Belskie created two life-size statues meant to represent the measurements of the "average" female and male American, based on government measurements of tens of thousands of "native white" men and women. They were named, without subtlety, Norma and Normman. Carved from white alabaster, with facial features and proportions approaching those of Greek gods, both figures embodied a particularly white, heterosexual ideal of what Americans should aspire to: "the perfect woman, the average American," was how Dickinson described Norma. After World War II, the figures would inspire contests to find which Ohio woman most resembled Norma—a difficult feat given her unnaturally high, rounded breasts.

Yet despite his eugenic-oriented views on what "normal" was, Dickinson rejected the idea that there was only one "normal" way for women to orgasm. In his practice, he began recommending to his married patients to stimulate the clitoris and labia during sex, including using the woman-on-top position to give the woman more control over clitoral contact. He once noted, without judgment, that one of his female patients had masturbated using an electric vibrator.

"Exalting vaginal orgasm while decrying clitoris satisfaction is found to beget much frustration," he wrote. "Orgasm is orgasm, however achieved."

————

Dickinson would never see the sexual revolution that would follow in his wake. In 1960, the FDA approved the Pill, the novel mix of synthetic estrogen and progesterone that promised to decouple reproduction from sex; within five years, 6.5 million women were on it. In 1965 the Supreme Court overturned the Comstock Law, a nineteenth-century federal statute that made it illegal to sell or distribute "obscene, lewd or lascivious" materials through the mail, including information about contraception. That decision upheld a woman's right to privacy, paving the way for *Roe v. Wade* to legalize abortion in 1973.

But he would help light the fuse. Dickinson was in his eighties and recently widowed when he met the man who would become his successor. Upon reviewing a research grant proposal from one Alfred Kinsey to analyze case studies of homosexual men, he was immediately struck by the scope of the younger man's research aims and methodology. Kinsey had started his career as an entomologist and professor of zoology at Indiana University, obsessively counting gall-wasps. After his students started bringing him their marital queries after class, he realized he had little real data to offer them, and pivoted to sex research. Dickinson decided that Kinsey was the right person to take on the mantle of sex research: objective, respectable, scientific. He bequeathed to him his blessing, his professional connections, and his 5,200 case studies before passing away in 1950.

Just as Dickinson had hoped, when Kinsey published his explosive study *Sexual Behavior in the Human Male* in 1948, he took pains to stress his objective, scientific lens. In contrast to Dickinson's goal of wedding "morals to normals," his goal was to leave morals completely out of it. As a bisexual man who grew up under the repressive thumb of his strict Methodist father, he knew how harmful it could be to have your proclivities judged by social mores. "Dr. Kinsey has studied sex phenomena of human beings as a biologist would examine biological phenomena," wrote a supporter at the Rockefeller Foundation, his main funding source, in the preface of his 1948 book. "The evidence he has secured is presented from the scientist's viewpoint, without moral bias or prejudice derived from current taboos."

But no amount of Rockefeller-sanctioned objectivity could prevent a book about human sexual behavior from raising the moral hackles of Americans. Even more controversial was the 1953 sequel, *Sexual Behavior of the Human Female*, for which Kinsey and colleagues statistically analyzed the sex histories of nearly 5,940 white American women. In dry, technical, but often shockingly graphic detail, they outlined the ways in which American society had gotten female sexuality wrong. Women were hardly chaste, virginal, and desiring only of their husbands in the missionary position, his survey findings suggested. Most masturbated frequently, and half of married women had had sex before betrothal.

Nor were they necessarily slower to arousal and orgasm, a truism often cited as a fundamental difference between the sexes. "It is true that the average female responds more slowly than the average male in coitus, but this seems to be due to the ineffectiveness of the usual coital techniques," he wrote—that is, hasty penetration by men. When left to their own devices, women had no problem experiencing orgasm. The most reliable route was masturbation; most women could easily achieve it in less than five minutes, usually by stimulating the labia and clitoris, rather than using penetration. Similarly, women who had sex with women had better sex, presumably because they understood their partner's anatomy better and focused on the clitoris—a lesson different-sex partners could profit from.*

---

* Specifically, they had more orgasms. In general, however, Kinsey challenged the commonly held idea that people were purely homosexual or heterosexual, pointing out that most people fell somewhere in between. His famous Kinsey Scale, first introduced in Sexual Behavior in the Human Male, provided six gradations between exclusive heterosexuality and exclusive homosexuality.

Kinsey had no patience for Freud and other psychoanalysts who had come up with their theories seemingly out of thin air, particularly when those theories were so clearly harming people's psyches and sense of self-worth. His book's foreword made the comparison explicit: Freud had theorized about women. Kinsey studied them. "Some of the psychoanalysts and some other clinicians insist that only vaginal stimulation and a 'vaginal orgasm' can provide a psychologically satisfactory culmination to the activity of a 'sexually mature' female," Kinsey concluded. "It is difficult, however, in the light of our present understanding of the anatomy and physiology of sexual response, to understand what can be meant by a 'vaginal orgasm.'"

To transfer one's orgasm from clitoris to vagina was not just preposterous—it was impossible. "There are no anatomic data to indicate that such a physical transformation has ever been observed," he wrote. This was unfortunate, because "some hundreds of the women in our own study and many thousands of the patients of certain clinicians have consequently been much disturbed by their failure to accomplish this biologic impossibility." One suspects Bonaparte certainly would have agreed.

Kinsey's findings would become a rallying point for the second-wave feminists and a jumping-off point for a sexual revolution. After he passed away, the sex researcher duo William Masters and Virginia Johnson would add evidence to his argument. In their 1966 book *Human Sexual Response*, they asked: was there any measurable difference between "vaginal orgasm" and "clitoral orgasm"? Armed with modern tools their predecessors couldn't dream of—including a plastic dildo-shaped camera equipped with fiber optics—they set about watching and filming couples in every stage of arousal. When they emerged, they had an answer. Orgasm, for both sexes, was a series of involuntary spasms, one every 0.8 seconds, involving increased blood flow to the genitals and higher muscle and nerve activity. It proceeded in four stages: arousal, plateau, orgasm, and resolution, which they illustrated with handy charts and graphs.

In the case of the female, they reported, those contractions might be felt in the outer vagina, the uterus, or the anus. But no matter where it was felt, the sensation always went back to the clitoris. Masters and Johnson declared this body part "a unique organ in the total of human anatomy" in that its only function was pleasure. Although "literature abounds with descriptions and discussions of vaginal as opposed to clitoral orgasms," from an anatomic viewpoint, they found no distinction. There was only one kind of orgasm, and it was clitoral.

A few days after her surgery, Aminata looked down at her clitoris and got a shock. It was huge, pink, and swollen. The next week, she started bleeding after standing up from the toilet, and rushed to emergency services. The doctor told her it was normal, nothing serious, but she was shaken. It was several weeks before the pain subsided, and her clitoris began to scab. "It's started to heal over; it almost doesn't hurt anymore," she told me on the day of her second follow-up appointment with Hatem, in November 2019.

By late December, her clitoris was finally beginning to turn the color of her skin. It was healing from the bottom, but the top was still a reddish-pink. The stitches were just beginning to fall out. But Aminata was already happier with her body. "Basically, I'm glad I'm not how I was before," she said.

It was still too early for Aminata to have felt anything pleasurable from her new anatomy. But Aïssa Edon, a thirty-eight-year-old woman who had her surgery with Foldès in 2005, vividly recalls the moment she first felt sensation in her clitoris. It was six months after her procedure, and she was wearing jeans for the first time, walking down the streets of Paris. Suddenly, she had a new sensation—like a spark. "I started to feel an electric shock down below," she recalls. She ran to the nearest toilet, pulled down her pants, and looked down at herself to see her clitoris engorged. "It was bulging," she recalls. "It was reacting to the friction of the jeans. It was quite crazy."

Aïssa was six when her stepmother took her to a local cutter in Mali, and twenty-three when she underwent reconstruction surgery in Paris. For her, surgery was never about sexuality. "It was about being complete," she says. "It was to have my body back, to have my power back." As she spoke, she used her hands to scoop air toward her chest, signifying the return of some intangible essence. The consequences that followed, she said, were just added benefits. She doesn't like to call her clitoris her "new" organ. But the sensation, she says, was new. "It was inside me but it was like, I couldn't visualize this organ, I couldn't see this organ. So obviously I had to relearn my body and understand how it works and everything, and who I am with this 'new' organ."

That relearning was about more than just anatomy. She learned to listen to her body, its desires, its hurts. She learned to trust her body, both by herself and with a partner. "Women, we are not really used to listening to ourselves, but African women even less," she says. After surgery, she began going to therapy to start to undo some of the assumptions she had absorbed about sex being bad, dirty, and shameful. "It was very important to understand I can think for myself," she says. "If I don't want to have sex with someone I don't need to have sex with someone. It'll be on my terms." Her experience led her to train as a midwife specializing in FGC, and ultimately, to become a midwife activity manager with Doctors Without Borders.

For Aïssa, the surgery was important because it gave her peace. "It's my journey, it's my story. And I will never force my journey on anybody else," she says. "But for me it gave me so much hope and so much more."

———

In 2010, Helen O'Connell saw another opportunity to make a contribution. As she began giving talks on the anatomy of the clitoris, she found herself peppered by questions from the audience. In particular, one journalist kept asking her about the G-spot. This "magic button," so well-established in the popular culture, had been described by some as a highly innervated part of the vagina, a zone that held the secret to powerful orgasm. Now this journalist was asking her: was it an anatomical reality, or a myth?

O'Connell wondered too. If it was real, why hadn't they come across these structures in their previous dissections? "I said, well, we never found it but we didn't look for it either. Which is a different scientific methodology," she says. "And so I decided we're going to set up something that is targeted toward establishing what's going on there."

The G-spot was first suggested by a man, but popularized by a woman. It is named for Ernst Gräfenberg, a Jewish-German gynecologist who invented the first modern intrauterine device while working in 1920s Berlin. In the 1930s, after the rise of the Nazi party restricted research on birth control, he fled Germany and eventually set up a gynecological practice in New York. There, like Dickinson, he found himself perturbed by how often the women he treated didn't reach orgasm during intercourse. "The solution of the problem would be better furthered," he wrote in 1950, "if the sexologists know exactly what they are talking about."

By this point, the idea of the clitoris as the control center of female pleasure had gained traction, along with the accompanying idea that the vagina was lacking in nerves. In Kinsey's studies, he had had five gynecologists test the sensitivity of the clitoris and other parts of genitalia by stroking these areas with a glass, metal, or cotton-tipped probe in nearly nine hundred women. While almost all women felt the sense of touch on their clitorises, only 14 percent were aware of their vaginas being touched. The vagina, he concluded, was an "insensitive orifice."

But Gräfenberg wouldn't give up on the vagina. He suggested that, for one particular portion of the vagina—a few centimeters up, on the belly side—there was actually something interesting going on. Like the male urethra, the female urethra is surrounded by erectile tissue that moves and swells during arousal. This "urethral erotic zone," he wrote, could be found in all women. In some cases, women who had orgasms through this zone released fluids "so profuse that a large towel has to be spread under the woman to prevent the bedsheets getting soiled." To him, this explained why so many women were left inorgasmic during intercourse: a lack of proper understanding of this magical region.

Gräfenberg's paper was overlooked in its time, and swiftly forgotten. And forgotten it would have remained, if it weren't for Beverly Whipple, a Rutgers University sexologist who was working as a nurse practitioner treating women with urinary incontinence. After finding that some women only leaked fluid during intercourse, she, too, located an unusual zone on the anterior wall of the vagina that seemed to swell with pleasure. In 1981, she appeared on the popular TV talk show *Donahue* to announce her findings: "If you stimulate this area, what happens, the orgasm occurs very rapidly, usually within a minute," she said, sitting next to two male physicians. "And people often report that they had many orgasms, frequently."

She added: "Missionary position just doesn't do it."

Donahue gaped.

She initially considered calling it the Whipple Tickle, at the tongue-in-cheek suggestion of a colleague. But at the time, there was already a well-known Mr. Whipple in a Charmin toilet paper commercial. The Gräfenberg spot it was (later shortened to the G-spot). She christened the area to honor Gräfenberg's contribution to female anatomy and pleasure. Then, she went on to co-write an entire book on the G-spot—the book that launched a thousand ships in search of one mythical spot. Real or not, Whipple certainly got the public to talk about female pleasure and sexual anatomy.

In the decade that followed, some researchers reported that the G-spot was a firm area of the vagina adjacent to the urethra, which could enlarge with stimulation. Meanwhile, others dismissed it as hypothetical, "an anatomical UFO," and even "a gynecological myth created for journalistic purposes." In 2012, a Florida gynecologist flew to Poland (presumably to avoid US regulations on cadaver research) and came back claiming to have found it in the fresh tissue of an eighty-three-year-old woman. It was, he said, a tiny sac of small, grapelike clusters of erectile tissue—"a deep, deep structure" nestled between layers of the vaginal wall. As it turns out, he was a plastic surgeon who advertised genital services including "G-spotplasty"—draw your own conclusions.

Though it was a compelling idea, the G-spot lacked data. And getting data was what O'Connell did best.

⟵

As she dissected more and more women—all in all, O'Connell estimates she's looked at fifty or sixty—she had become particularly fascinated by the two bulbs of the clitoris. Like flattened tulip bulbs, they hugged the vaginal walls, and were remarkably dynamic.* Encased only in a gossamer-thin, elastic lining, they could grow and change vastly in shape.

In anatomy books, they were usually colored a different hue from the clitoral glans and arms, or crura, and were still labeled "bulbs of the vestibule." Kobelt, she noted, with a typical focus on penetrative sex, had asserted that their main function was to squeeze the penis. "With the stiffening of its walls . . . it transforms itself into a sucking tube . . . The sponge-like elastic padding of the vaginal tube added to the bulbi at the entrance, qualifies the vagina splendidly for its main purpose; for a gentle and yet close grasping," he wrote.

But to O'Connell, they quite obviously belonged to the clitoris. They connected to the body of the clitoris at a root, forming a mass of sensitive erectile tissue. They were also made of a similar spongy erectile

---

* Kobelt, by contrast, described each bulb as having "the shape of a tremendously distended leech."

tissue to the body and crura. Together, the bulbs and arms were exactly like the erectile pillars of the penis, just splayed out in space instead of together in one column. What made her so confident? "Well, I'll show you," she says, flipping through the pages of her thesis. She pauses at a photograph of the bulbs with the pelvic bone removed, revealing how they connect to the root of the clitoris along with the crura.

"In terms of 'Is this a unit, are these things related or unrelated?' you can make up your mind," she says. But "the idea that you're going to say, 'Oh, that belongs over there to the vestibule, rather than as part of this unitary thing,' seems absurd. And not helpful. *Not helpful.*" The main difference, to her, is that the bulbs have the freedom to expand. The crura are surrounded by a *tunica albuginea* (Latin for "white coat")—a thick, fibrous sheath that prevents them from going anywhere.* But the bulbs could expand outward, causing the labia to puff and swell— "make them red?" she suggests, a sly smile on her lips. We don't know yet, because no one has done the functional studies. She wishes somebody would.

Maybe these bulbs held the key to the so-called G-spot. In 2017, O'Connell and her colleagues carefully dissected the urethras and vaginal walls of thirteen female cadavers, ranging from thirty-two to ninety-seven years old. Then, they used MRI imaging to probe the area further. Besides the clitoris, they found no other erectile or spongy tissue. Instead, the G-spot seemed to press up against the part of the urethra where the body, arms, and bulbs of the clitoris all combine. Although the vagina itself contained no erectile tissue, pressure on it can be felt on the aroused internal clitoris, particularly the bulbs. For some women, this sensation can be intense, leading some anatomists to dub it a unique spot. The G-spot, she concluded, "does not exist as an anatomic construct."

As it turns out, the G-spot had become a phenomenon a bit like

---

* Interestingly, this makes an old term for the clitoris, *tentigo* or tension, somewhat accurate.

the vaginal orgasm: a near-mythical idea that made women feel pres-
sured to have a certain type of sexual experience, while reassuring men
that they had to do very little in bed. It made female sexuality seem
straightforward, achievable. "There is this mythological kind of desire
for a magic button in the vagina, that if you just do 'uh'"—O'Connell
makes a low grunting noise—"it's all going to be fine. You know?
Don't actually worry about getting to know this woman, you've just
got the magic wand and you're going to be able to—with the right
type of thrusting and her doing exactly the right thing—it's all going
to work simultaneously."

Similarly, the "vaginal orgasm" assured men that whatever they were
doing in bed was just fine. But to actually work, she explains, male
thrusting would have to indirectly stimulate the clitoral bulbs. That's
not impossible—"it's not like it's on another planet," she says—"it's just
such an indirect approach, isn't it?"* In her thesis, she uses her finger
to trace the outline of the clitoris in relation to the anatomy around it,
showing how it interacts with the urethra and vagina. "You can see the
boomerang shape of the clitoris right there," she says. "The fact that
you're going to ignore this tissue in your plan . . . yeah, that's pretty
unlikely to work."

The G-spot and the vaginal orgasm were both based on a limited
concept of female pleasure as being either/or, clitoris versus vagina.
Unfortunately, "it doesn't work like that," O'Connell says with a laugh.
"It doesn't matter how much you want it to work like that, it won't."

———

As it turns out, the founder of the G-spot isn't pleased with how her
idea had been taken by the world, either. Whipple, now eighty-seven,

---

* Surveys find that between 20 and 30 percent of women have the ability to orgasm
from penetration alone—not many, but some. This phenomenon, like the G-spot, can
be explained by the stimulation of the clitoris through the vaginal walls.

never wanted to create the hunt for the magic button. Her intention was simply to validate women's lived experiences with research—to make them feel normal, and let them know they weren't alone. Forty years later, she says she regrets the way the media used her concept to make it sound like the anatomy of female orgasm was so simple. The G-spot wasn't just an on/off switch or a single button, she stressed. In her later work, she explained that it was a complex junction of tissues, including glands and the erectile tissue of the clitoris.

But nobody seemed to want to listen to that. They all wanted the magic button. "I guess we misled people, because it's more than one little spot, it's a whole area," she said in 2016, on a podcast called *Science Vs.* "But at that point, you know, that's what we called it." You can't put the demons back in Pandora's box.

Whipple still believes that this part of the body should be named after Gräfenberg. But if she could go back in time, she would have called it an "area" instead of a "spot." "I'd like women today to know that we are all capable of many pleasurable, sensual, and sexual experiences," she told me in 2021, over Zoom. "That it doesn't just rely on a G-spot or a clitoris. That there's so much more to our bodies." These days, she focuses on the diffuse pleasure felt in every part of the body. To this end, she's created an extragenital matrix, a grid listing thirty-five body parts and types of touches meant to help couples figure out what gives them pleasure. "Sometimes holding hands, or touching, whatever it is, that feels good to you, is an end in itself," she said on the 2016 podcast.

Personally, she added, she loves her husband to suck on her big toe, "but we're all different." The problem with the G-spot was it made us sound like we were all the same.

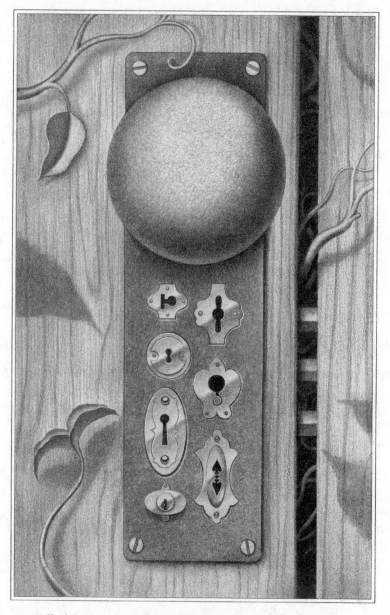

Full of glands and muscles and collagen and changing constantly and
fighting pathogens all the time . . . It's just a really amazing structure.

—PATTY BRENNAN

CHAPTER 3

Resilience

(VAGINA)

D r. Patty Brennan never intended to become a champion of the vagina. Her journey, in fact, began with a penis.

It was a late summer afternoon in 2000, and the twenty-eight-year-old Colombian biologist was stalking her study animal, a squat gray-blue bird called the great tinamou (*tin-ah-moo*), in the dense Costa Rican rainforest. As always, the forest floor was dark and shadowy, the sunlight swallowed up by the upper canopy. It was stiflingly humid; she was sweating through her protective gear. All around her was the unending buzz of mosquitoes and the mournful call of birds. "You could die in that forest, and there would be no trace of you in just a few months," Brennan recalls. "You would disappear completely."

That's when she heard it: a pure, whistling tone. A male tinamou, calling for a mate. As she held her breath, a female appeared from the dense underbrush. She ran up to him, backed away, then chased him again. Finally she crouched down with her tail in the air, inviting him to mount. As Brennan watched through her binoculars, the male clambered clumsily onto her back. Splaying her wings to the side—"it looked so uncomfortable for this poor female," she recalls, wincing—he grabbed her neck with his beak to stabilize himself.

Brennan will never forget what happened next.

For most birds, mating is an artless affair. That's because they don't have external genitalia, just a multipurpose opening under the tail used to expel waste, have sex, and lay eggs.* They briefly rub genitals together in an act known as a "cloacal kiss," in which the male transfers sperm into the female. The whole event takes seconds. But this time, the pair began waddling around, glued together. The male started thrusting. When he finally detached, she saw something dangling off him—something long, white, and curly.

"What the hell is that thing?" she recalls thinking. "Oh, God, he's got worms."

Then she had another thought: "Man, is that a penis?"

Birds, she thought at the time, didn't have penises. In her two years studying them at New York's Cornell University, a world leader in avian research, she'd never once heard her colleagues mention a bird penis. And anyway, this certainly didn't look like any penis she had ever seen—it was ghostly white, curled up like a corkscrew, thin as a piece of cooked spaghetti. Why would such an organ have evolved, only to have been lost in almost all birds? That would have been "the weirdest evolutionary thing," she says.

When she returned to Cornell, she decided to learn everything there was to know about bird penises—which turned out to be not so much. Ninety-seven percent of all bird species have no phallus. Those that did, including ostriches, emus, and kiwis, sported organs quite different from the mammalian variety. Corkscrew-shaped, they

---

* Biologists usually call this orifice a cloaca, which means "sewer" or "drain" in Latin. Brennan simply refers to it as the vagina, since it performs all the same functions and then some.

exploded out into the female in one burst, and engorged with lymphatic fluid rather than blood. Sperm traveled down spiraling grooves along the outside.

Brennan, an animated woman with an impish smile and no qualms about getting a few intestinal parasites on her iPhone while dissecting a snake, had been the first to observe a penetrative penis in this particular species of tinamou. Only later would she ask the question that would distinguish her from all her peers: If this was the penis—then what were the vaginas doing? "Obviously you can't have something like that without some place to put it in," she would tell the *New York Times* in 2007. "You need a garage to park the car." For the first time, she wondered about the size, shape, and function of that, er, garage.

———

Biologists love penises. And with good reason: they're some of the most wildly varying organs in the animal kingdom. There are penises that can taste, smell, and sing. Ones that look like corkscrews, crowbars, and glowing blue lightsabers. A penis can stretch up to nine times your body length (if you're a barnacle); be a detachable tentacle covered in suckers (if you're an argonaut octopus); or even *see,* using light-sensing cells that guide it to its destination (if you're a Japanese yellow swallowtail butterfly). They're also easy to study. Penises kind of just hang out, in contrast to the more internal female apparatus.

But the reason for centuries of vaginal neglect goes deeper than that. Historically, researchers have long considered the vagina the more-passive, less-important half of the genital equation. When Freud chose the vagina as his symbol of proper womanhood, he was drawing on what we knew from biology. This organ, he believed, had one job: to receive the penis. Its name reflects this: coined by Italian anatomist Gabriele Falloppio, "vagina" derives from the Latin for "scabbard," "sheath," or "close covering," suggesting that its main function is to house a sword. (Although, oddly, the word penis comes from the Latin for "tail.") For women to center their sexual desire on the vagina, therefore, meant to accept their role as penis recipients and baby-makers.

When the second-wave feminists challenged Freud's ideas, they did so by lambasting the vagina. The vaginal orgasm, they argued, was a concept invented by Freud and perpetuated by men to maintain control over women's sexuality. "Actually the vagina is not a highly sensitive area and is not constructed to achieve orgasm," said New York feminist Anne Koedt in her 1970 speech *Myth of the Vaginal Orgasm.* "It is the clitoris which is the center of sexual sensitivity and which is the female equivalent of the penis."[*] Her words helped elevate the clitoris into a symbol of female independence and liberation. But in doing so, they inadvertently helped denigrate the vagina as dull and not worth writing home about.

Animal researchers, too, have long dismissed the vagina. Historically, entomologists used insect penises as key features to describe species and tell them apart—meaning they tended to preserve, study, and illustrate male structures while ignoring female ones. When preparing pairs of insects mid-copulation, they often performed a step called "clearing," in which they essentially washed away the female reproductive tract in order to better view her other parts, as journalist and biologist Emily Willingham learned while writing her 2020 book, *Phallacy: Life Lessons from the Animal Penis.*[†]

[*] In the same speech, Koedt deplored Marie Bonaparte as one of the Freudians who had drunk the patriarchal Kool-Aid and "ran around absurdly trying to change female anatomy to fit their basic assumptions."

[†] Similarly, Dr. Frances Conley, the first female professor of neurology in the United States, recalled that in her dissection training at Stanford Medical School in the 1960s, "breasts of female cadavers were an unnecessary appendage summarily removed with

Willingham knew from graduate biology classes that males were better studied. "But I hadn't realized the extent to which [the female reproductive tract] had been just deliberately, explicitly dismissed," she says. Not only was that biased, she realized—it was unscientific. Genitals are structures that shape each other, in a mutual dance that plays out over generations. "These are two extremely closely involved structures that are also closely involved in reproduction. I mean, it's a direct line to fitness," she says. How could science have missed something so obvious?

Today, researchers like Brennan are finding that vaginas are more complex and variable than anyone thought. Rather than passive, they often play active roles in deciding whether to allow intruders in, what to do with sperm, and whether to help a male along in his quest to inseminate. From dolphins to snakes to alpacas to bats, Brennan is finally opening what she calls "the copulatory black box" of female genitalia. To her, the vagina is a remarkable organ in its own right, "full of glands and full of muscles and collagen, and changing constantly and fighting pathogens all the time," she says. "It's just a really amazing structure."

So far, she is one of few scientists exploring this territory. "It's an exercise in both excitement and frustration, because every time I get into a different system and I start looking at questions about why vaginas are the way they are, I end up realizing that nobody really knows," she explains. "And I'm going to have to keep plowing my way into answering these questions for the first time."

But before we get to that, we have to make a pit stop back at penises.

⌐

In 2005, the pursuit of penises led Brennan to the University of Sheffield, a sprawling campus in the English countryside surrounded by farmland. After realizing that "there is a huge gaping hole in our knowl-

---

rapid scalpel strokes in order to reveal the musculature of the anterior chest wall, which was then studied in intricate detail."

edge of this very fundamental part of bird biology," she had pivoted her research and was now a postdoctoral researcher focusing on bird-penis evolution. She was here to learn the art of dissecting bird genitalia from Dr. Tim Birkhead, an evolutionary ornithologist who studied sperm competition. She got to work dissecting quail and finches, which had little in the way of outward genitalia. Next, she opened up a male duck from a nearby farm, and gasped.

The tinamou's penis had been thin, like spaghetti. This one was thick, and massive. Whoa, she thought. Wait a minute—where is this thing gonna go?

No one seemed to have an answer. The problem was, the typical bird-dissection technique focused almost entirely on the male. When researchers did dissect a female duck, they sliced all the way up through the sides of the vagina to get at the sperm-storage tubules near the uterus (in birds, it's called the shell gland), distorting their true anatomy. They tossed the rest out, unexamined. When she asked Birkhead what the inside of a female duck looked like, she recalls, he assumed it was the same as any other bird: a simple tube.

But she knew there was no way an appendage as complex and unusual as the duck penis would have evolved on its own. If the penis were a long corkscrew, the vagina ought to be an equally complex structure.

The first step was to find some lady ducks. Brennan and her husband drove out to one of the surrounding farms and purchased two Pekin ducks, which she euthanized without ceremony on a bale of hay.* Instead of slicing the reproductive tract up the sides, she started peeling away the tissues, layer by layer. She labored over the vagina, carefully pulling away the layers of connective tissue that surrounded it, "like unwrapping a present," she says. Eventually, a complex shape emerged: twisted and mazelike, with blind alleys and hidden compart-

---

* Brennan's husband is used to these kind of excursions. "He brings me roadkill as a nuptial gift," Brennan says with a laugh. "He knows me well."

ments. But there was something even more odd: its spirals twisted in the *opposite* direction of the male's.

When she showed Birkhead, they both did a double-take. He had never seen anything like it. Could it be a fluke? But when she dissected the second duck, she found the same thing. Then Birkhead called a colleague in France, a world expert on duck reproductive anatomy, and asked him if he'd ever heard of these structures. He hadn't. The colleague went to examine one of his own female specimens, and reported back the same thing: an "extraordinary vagina."

"The secret was in the dissection," Birkhead and Brennan would write. "The pouches were bound to the vagina wall by a dense matrix of connective tissue which also concealed the spiral, so that on casual inspection, the pouches and the spiral were both invisible. By carefully dissecting away the connective tissue we exposed one of the most extraordinary secrets of the female duck's biology."

It seemed like the females were responding in some way to the males—and vice versa. But the vagina didn't look like it had evolved to accommodate the penis. It looked like it had evolved to evade it. "I couldn't wrap my head around it. I just couldn't," Brennan recalls. She preserved the structures in jars of formaldehyde and spent days turning them over, trying to figure out what could explain their complexity.

That's when she began thinking about conflict. Duck sex, she knew, could be notoriously violent. Ducks tended to mate for at least a season. However, extra males lurked in the wings, ready to harass and mount any paired female they could get their hands on—"a case of rape by act," as Darwin might have said. This often leads to a violent struggle, in which males injure or even drown the female. In some species, up to 40 percent of all matings are forced. The tension is thought to stem from the two sexes' competing goals: The male duck wants to sire as many offspring as possible, while the female duck wants to choose the father of her children. Sometimes, males seek to overpower female resistance by force.

This story of conflict, Brennan suspected, might also shape genitalia. "That was the part where I was like: holy cow," she says. "If that's
really going on, this is nuts." She started contacting scientists across
North and South America to collect more specimens. One was Kevin
McCracken, a geneticist at the University of Alaska who, while out
on a wintry jaunt, had fortuitously discovered the longest known bird
phallus on the Argentine lake duck, which unraveled to a stunning
17 inches. He had suggested that perhaps the male was responding to
female preference—wink-wink, nudge-nudge—but hadn't bothered
to actually examine the female.

When Brennan called him up, he was more than happy to help her
collect more specimens. Today, he admits that perhaps the reason he
hadn't considered looking at the female side of things was a result of his
own male bias. "It was fitting that a woman followed this up," he says.
"We didn't need a man to do it."*

Ultimately, Brennan analyzed the vaginas of sixteen species of
waterfowl to find remarkable diversity compared to what we knew of
any other bird group. There was a lot going on inside those vaginas.
The main purpose, it appeared, was to make the male's job harder: It
was like a medieval chastity belt, built to thwart the male's explosive
aim. In some cases, the female genital tract spiraled the opposite direction of the penis, so that the penis couldn't fully inflate; it was full of
pockets where sperm went to die. There were also muscles surrounding
her cloaca that could block an unwanted male, or dilate to allow entry
to a preferred suitor. This prevented males from fully entering without
female cooperation.

The correlation between penis size and shape, and vagina size and
shape, was "incredibly tight," Brennan told the *New York Times*. "When

---

* After we spoke, McCracken emailed me a photo of the infamous lake duck member. "I got to thinking since you are writing a book about the vagina, why not have
a photo of a lopped-off penis to go with it?" he wrote. It looked like a deflated,
flesh-colored Slinky in a jar.

you dissected one of the birds, it was really easy to predict what the other sex was going to look like."

Whatever the females were doing, they were succeeding. Brennan found that in ducks, only 2 to 5 percent of offspring are the result of forced encounters. The more aggressive and better endowed the male, the longer and more complex the female reproductive tract became to evade it. It was a struggle for reproductive control, not bodily autonomy: Although a female couldn't avoid physical harm, her anatomy could help her gain control over the genes of her offspring after a forced mating. The vagina, Brennan realized, was by no means passive or simple. This one, as Willingham put it in *Phallacy*, was an expertly rigged penis-rejection machine.

Before that, reproductive biologists had "assumed that female reproductive anatomy was just so much plumbing that the male's sperm had to traverse," Brennan wrote. "In other words, females were assumed to be passive participants in reproduction." Yet at least in waterfowl, both sexes had clearly evolved together. The story was all there, written in genitalia.

A world opened up before Brennan's eyes: the vast variety of animal vaginas, wonderfully varied and woefully unexplored. For generations, anatomists had praised the penis, fawning over its length, girth, and weaponry. Brennan's contribution, simple as it may seem, was to look at both the male and female genitals—and how they worked together. "I'm like, Really, am I that lucky that I like stumbled on this crazy adaptation right away?" she says. "And then I was like, Nobody's that lucky. I bet you this is more common than we know, it's just that nobody's looking. And that's why I decided to look everywhere."

And why hadn't anyone bothered to look? Perhaps, for the same reasons Charles Darwin hadn't.

In 1879, Darwin got himself into hot water over the hindquarters of some monkeys. The aging naturalist had drawn a huffy remark from a critic that he would spend his time more usefully bottling air than contemplating "the hinder parts of monkeys." The comment was in reference to Darwin's writing on the "indecorous habit" that certain simians had of showing off their crimson buttocks. Amused, Darwin wrote to a friend that it was "very acute of Mr. Ruskin to know that I feel a deep & tender interest about the brightly coloured hinder half of certain monkeys."

What lay behind Darwin's fascination with these bold monkey bottoms was sexual selection. In the second half of his career, he proposed another force that shaped species in addition to his well-known theory of natural selection. That force was, essentially, female preference. See, in order to pass on one's genes, one has to mate.[*] And in order to mate, one has to appeal to the opposite sex. Therefore, males evolved elaborate traits—the choruses of frogs, the colorful plumage of birds—not for any adaptive reason but merely to please the females doing the selecting. Like the peacock's tail, Darwin speculated, the male monkey's swollen rump was a sexual ornament that he displayed to attract the female during courtship. He noted that "these parts are more brightly coloured in one sex than the other" and that they "become more brilliant during the season of love."

Odd, then, that it wasn't just males doing the displaying. In his less-iconic second work, *Descent of Man*, he wrote that it was the female Rhesus monkey whose naked skin around the tail was "of a brilliant carmine red" and "periodically becomes even yet more vivid." During the mating season, her vulva—the "adjoining parts," as he would later delicately put it—swelled with blood, signaling her readiness to mate. Similarly, he once documented a female crested black macaque who, "with gurgling tones, turned around and showed the male its red rear

---

[*] Usually; see Chapter 5.

end, which I had never seen from this animal before. At this sight the male got excited, because he beat violently on the bars, all the while gurgling loudly."*

The male Rhesus, meanwhile, had no trace of red. This fact didn't fit his thesis, so he only mentioned it in passing, and comforted himself with the fact that in size, teeth, and whiskers, the male still "follows the common rule of the male excelling the female."

It was about the closest Darwin would get to mentioning a vulva or vagina. Despite his book's full title—*The Descent of Man, And Selection in Relation to Sex*—he took pains never to address primary sex characteristics, aka genitalia. Genitals "do not here concern us," he wrote. He considered them functional rather than ornamental, and therefore not subject to forces of sexual selection. Instead, he limited his scope to features that seemed to attract a female to mate in the first place, "such as the greater size, strength, and pugnacity of the male, his weapons of offence or means of defence against rivals, his gaudy colouring and various ornaments, his power of song, and other such characters."†

Darwin knew what he could and couldn't get away with. As a young man, he had waxed poetic on the barnacle penis, which, at eight to nine times the length of its owner, was one of the longest in the animal kingdom. In four lengthy barnacle-themed volumes—more than 1,200 pages—he called it "wonderfully developed" lying "coiled up, like a great worm."‡ But these days, he was no longer writing only

---

* He wrote this bit in German, a technique he often used to prevent scandalizing the ladies.

† For instance, beards. He wrote that human ancestors "acquired their beards as an ornament to charm or excite the opposite sex." Darwin grew his own iconic face hedge out at age fifty-three, at the behest of his wife.

‡ Barnacles are usually immobile and almost always hermaphrodites, meaning they have both male and female sex organs. The reason they have penises at all is so they can reach across the divide and squirt sperm at other barnacles, instead of resorting to self-fertilization.

for an audience of barnacle buffs. Now he was a public figure, gray of beard and noble of gaze. Moreover, he was on a mission that already amounted to blasphemy: in *Origin*, he postulated that all animals on Earth, rather than being shaped by a divine hand, had come into being by a series of random accidents and natural laws. Now he had dragged humans down into the muck.

Given this delicate task, he had to be cautious. "Darwinians had to be like Caesar's wife: above suspicion," says Evelleen Richards, a historian of science and author of *Darwin and the Making of Sexual Selection*. "They had to present themselves as very, very respectable gentlemen because their opinions were so controversial." He could hardly afford to bring penises and vaginas into it.

Read between the lines, however, and there was one other exception. One of the biggest conceptual leaps Darwin made in *Descent* was arguing that sexual selection was powerful enough to explain the variation among human races, from skin color to genitalia. As a key piece of evidence, he cited reports by European colonists on the unusual backsides of KhoiKhoi women from the Cape of South Africa. In these women, Darwin wrote that "the posterior part of the body projects in a wonderful manner" and added in a footnote (in Latin) that "the very girdle or protuberance on women which we see as repulsive is thought to be of considerable value by the men of this tribe." This "girdle" was a covert reference to the elongated labia minora that hung down in some women, which had long attracted the prurient attention of European anatomists.*

Why, when Darwin took such pains to avoid talking about genitalia, did he linger on the vulvas of monkeys and KhoiKhoi women? "Because they were not seen as human," says Dr. Banu Subramaniam, a professor of women, gender, and sexuality studies at the University of Massachusetts, Amherst, and author of *Ghost Stories for Darwin: The Science*

---

* His "expert" source, by the way, was a retired army surgeon who had been stationed in South Africa thirty years earlier and hadn't set foot on the continent since.

of Variation and the Politics of Diversity. In Darwin's eyes, both were pre-human, primitive ancestors to today's refined white European women. Commenting on their sexual anatomy did not come with the risks that talking about proper Victorian women did.

Like skull size and skin color, the sexual traits of South African women were taken as evidence that they fell at the bottom of the human hierarchy—a convenient notion for a country still imposing imperial rule. "Sex and race are intertwined," Subramaniam says. "Sex is always raced, and race sexed. These views remain with us in, say, the image of the hypersexualized Black woman. To me this is why the history of science is so important: it helps us understand the pernicious legacies of slavery that live on in the racism of today."

In the 1830s, when Darwin was sailing around the world in the HMS *Beagle*, British women could not vote, go to university, or own property. Women and men were considered distinctly unequal—two complementary halves of a civilized whole. By the time he wrote *Descent*, norms were changing. British suffragists were demanding the vote and vying for access to higher education and the professions. The National Society for Women's Suffrage had recently been established. Eminent thinkers like the philosopher John Stuart Mill were beginning to challenge the accepted wisdom that women were naturally inferior. Perhaps, Mill argued in his book *The Subjection of Women*, woman's current status was not indicative of her innate potential but of her social conditions. Treat women as equals, and the perceived distinction would disappear.

Darwin, a product and pillar of an earlier time, disagreed. To him, a woman was meant to be—as he had once written in his journal as a young man contemplating marriage—"a nice soft wife" and "an object to be beloved and played with. Better than a dog anyhow." (She should also, importantly, have money.) Soon after Mill's book was published, he ran into a woman who had reviewed it, the passionate British feminist and animal rights activist Frances Power Cobbe, while on a stroll in the Welsh countryside. Standing sixty feet above her on the hillside path, he told her exactly what he thought about the position of women. "Mill could learn some things from physical science," he barked down. "It is in the struggle for existence and (especially) for the possession of women that men acquire their vigor and courage." *

Cobbe listened patiently, then offered him a copy of Immanuel Kant on "moral sense," which he declined.

Darwin's attitudes toward women were a direct outgrowth of his ideas about other animals—or, rather, each reinforced the other. Throughout his career, he insisted that female animals were less capable and intelligent than the male of the species. In nearly every species, "it is the males that fight together and sedulously display their charms before the females; and those which are victorious transmit their superiority to their male offspring." This, in his view, was why males tended to be "bolder and fiercer" and to grow fantastically elaborate traits, like the peacock's magnificent tail. The same logic held with humanity: "Thus man has ultimately become superior to woman," he concluded.†

That left females as the duller, chaste sex. For Darwin, the female occupied a lack: a lack of horns, a lack of luminescence, a lack of beauty, a lack of brains. To him, she was little more than a shadow, a foil next to which the male appeared all the brighter. "Women have served all

---

* He was also strenuously against birth control. If only educated and well-bred women had access to contraception, he believed, they would soon be overtaken by the poor and downtrodden masses—inferior, "reckless" types of people. Birth control "would spread to unmarried women and would destroy chastity, on which the family bond depends; and the weakening of this bond would be the greatest of all possible evils to mankind," he warned. Darwin certainly followed his own maxim: he and his wife Emma had ten children, seven of whom survived to adulthood.

† He did admit, as a consolation prize, that "women have become more beautiful."

these centuries as looking-glasses possessing the magic and delicious power of reflecting the figure of man at twice its natural size," Virginia Woolf would write decades later. "That is why Napoleon and Mussolini both insist so emphatically upon the inferiority of women, for if they were not inferior, they would cease to enlarge."

His argument that women were built for reproduction and men for loftier aims would be used to keep women out of the academy for almost a century. In Darwin's time, higher education was thought by some authorities to shrivel a woman's ovaries and keep her from her motherly duties. In 1873, Harvard medical professor Edward Clarke wrote that he had seen women who "graduated from school or college excellent scholars, but with undeveloped ovaries. Later they married, and were sterile." During World War II, when Harvard Medical School debated whether to accept women, one faculty member argued that doing so would violate "the fundamental biological law that the primary function of women is to bear and raise children."

Like Freud, Darwin had a blind spot when it came to women. Although the ideas that would earn him eternal recognition were based on a theory of gradual change, in his mind, women were somehow both unchanging and unchangeable. Once, while attempting to explain the development of religion, he mused that human beings have a biological need to believe in something "other." Scientists, too, it seemed, needed an other—a different, inferior type of human, one whose existence reinforced and maintained their own superiority. For Darwin, the differences between the sexes were bedrock. Woman's inferiority was immutable. It was written into her biology, woven into her flesh.

And so, in a Darwinian world where women were uniformly passive and dull, there was no need for scientists to investigate their sexual orifices. They already knew what they would find: passive and dull. QED. End of story.

Or was it?

The field would wait another hundred years to find out.

In the century that followed, biologists had few tools to explain the astonishing spectrum of genital diversity they found. They often relied on the explanation of "lock and key," a Darwinian maxim that stated that genitalia were species-specific, so that you didn't accidentally stick your key in the wrong lock, or vice versa. It was nature's form of two-step authentication, a way of telling you that maybe it wasn't such a good idea to proceed.

But that explanation could only go so far. At least this was the suspicion of William Eberhard, an entomologist at the Smithsonian Tropical Research Institute in Costa Rica. Eberhard had started out studying spiders, a class of animals in which female genitals are partially outside the body and made of hard exoskeleton, making them more obvious to researchers. When males and females interact, much of those interactions take place outside the body. He knew that male genitals varied widely even among closely related species and were often more elaborate than they should be simply for appendages that transferred sperm.

Sexual selection, he realized, didn't stop at the decision to mate. Somehow, in a major oversight, Darwin had failed to consider that not all copulations lead to reproduction. At least part of the outcome depended on the complex dialogue between males and females during copulation. "It is impressive," he wrote, "to see the long shadow that one of Darwin's few omissions had in the history of studies of sexual selection."

Males, he noticed, worked hard to please and stimulate their mates during sex, something they wouldn't have bothered with if they were already guaranteed paternity. What if penises didn't just transfer sperm but were also courtship devices? "All those males that were so busy rubbing, squeezing, shaking, tapping, feeding, etc. their mates were telling me something," he wrote me in an email, "that they needed to induce positive responses in their females, even though they had already convinced them to copulate." In 1985, he delved into alternative reasons for genital evolution in the book *Sexual Selection and Animal Genitalia*.

Unfortunately, that book left readers with the same impression as Darwin: that females are less variable, vaginas less interesting, and penises were the star of the show. (In one instance, he referred to males as the players, and females as merely "the field on which males compete.") Penis studies flourished, while vaginas continued to fall by the wayside. But Eberhard ultimately realized that he needed to understand what was going on inside the female body. "The idea that perhaps male genitalia were courtship devices set me to thinking about what the function of such courtship might be—and only then did I begin to see how many crucial female-controlled processes stand between a male and siring her offspring even after he has begun to copulate with her," he wrote in an email.

Leafing through the literature, Eberhard concluded that even as sperm competed within the female body, the female had tricks up her sleeve to influence the race. In his second book, *Female Control,* he outlined the myriad strategies that the female had for wresting control over the genetics of their offspring from the male. Many were made possible by her remarkable reproductive anatomy: vaginas that stored, rejected, or destroyed unwanted sperm. They included:

Sometimes Discard Sperm of Current Male

Sometimes Prevent Complete Intromission and Ejaculation

Sometimes Fail to Transport Sperm to Storage Organs or Fertilization Sites

Sometimes Remate with Another Male

Sometimes Fail to Ovulate

Sometimes Fail to Prepare Uterus for Embryo Implantation

Sometimes Abort Zygotes

Choose Among Sperm That Have Reached the Egg

Sometimes Forcefully Terminate Copulation before Sperm Are
   Transferred

He called these strategies "cryptic female choice"—cryptic because they were invisible to scientists, as well as to a female's mate. Even after copulation, he concluded, the female remained firmly in control.

Eberhard's work would open the door to a wide array of papers on female genital evolution. (Though, he regrets, few on humans.) Today, he marvels at how long it took him to see what the field was missing. "Our unconscious biases surely have strong effects on the questions we ask and even what we are conscious of seeing," he wrote in an email. "I know that I had exactly the same bias; earlier, before I started thinking about genital evolution and [cryptic female choice], I watched animals perform copulatory courtship and didn't even write it down in my notebook, much less wonder why it was occurring."

Once again, the maxim holds: we can't see what we aren't looking for.

————

There may be some truth to the idea that females are less variable than males, says Brennan. Here's why: Scientists define a penis as any appendage used to transfer sperm. That means that almost any body part—an arm, a fin, a tentacle, an antennae—can be adapted by evolution for this purpose.

Meanwhile, the female already has a structure in place for sperm uptake: the birth canal, a tube leading from the uterus to the outside. Long before penises hit the evolutionary scene, this tube already had its primary function, which was to eject eggs. It's since become the ultimate multitasker. "Female genital tracts are under multiple sources of selection: not just mating, but also storing sperm, egg laying, birthing, and often interfacing with the terminal portion of the digestive tract," wrote Brennan in 2015. And this multifaceted tube can only vary so much while keeping its other functions.*

So yes, the female has some extra constraints on her. If the male is a free-verse poem, she's more of a sonnet, limited by certain formal rules. But that doesn't mean you can't have immense variation within those constraints. (See: "Shall I compare thee to a summer's day?" by William Shakespeare versus "To Time" by Sylvia Plath.) The problem is that female genitals had long been assumed to be more static than the male, and as a result, not as well studied, perpetuating that assumption.

In 2014, a meta-analysis on gender bias in genitalia studies spelled out the problem. The authors analyzed 364 papers on genitalia diversity to find that investigators had long focused disproportionately on males—and that this bias was growing stronger. Most simply overlooked the female side, while others relied on mathematical models to extrapolate out what the female might look like. "We argue that the persisting male bias in this field cannot solely be explained by anatomical sex differences influencing accessibility," wrote the authors. "Rather the bias reflects enduring assumptions about the dominant role of males in sex, and invariant female genitalia."

Vaginas, indeed, did suffer a lack. Not a lack of beauty, strength, or vigor—but a lack of knowledge, data, and curiosity. "Understanding genital evolution," the authors concluded, "is hampered by an outdated single-sex bias."

---

* Notable exceptions include koalas and kangaroos, which have a rare three-vagina configuration in which penis entry and birth are decoupled. They have two tubes for insemination and one down the middle for birthing—meaning they can, unfortunately, be constantly pregnant.

In other words, there was a big question mark in the literature about what, exactly, was going on in all those vaginas.

———

In March 2013, Brennan was in her office at the University of Massachusetts, Amherst, checking her daily Google Alert for "duck genitalia." An article popped up from a conservative website called CNS News, which described its mission as countering "liberal bias" in media. They had identified what they described as the ultimate waste of taxpayer money: a National Science Foundation grant for $384,949 to Yale University for research on duck penises.

At first, Brennan ignored it. But before long she had no choice but to pay attention—the headline was popping up on every newspaper, every website. "It's part of President Obama's stimulus plan, and it's just one example of the kind of spending decisions that have added up to massive debt and deficits," Fox news anchor Shannon Bream announced on air. Fox's website posted a survey that asked, "Was duck penis study an appropriate use of taxpayer money?" (The most common answer: "No—what a quack!") Soon *Mother Jones* had dubbed the media cyclone "Duckpenisgate."

Brennan was aghast at how her work had been taken out of context. "It was the worst time in my entire scientific career," she recalls. "I wanted to crawl under my desk and never emerge again." Eventually, she decided that she couldn't just sit there—she needed to enter the conversation and help educate the public. Against her colleagues' advice, she wrote an editorial in *Slate* advocating for the value of basic science. "Genitalia, dear readers, are where the rubber meets the road, evolutionarily," she wrote. "To fully understand why some individuals are more successful than others during reproduction, there may be no better place to look."

After making a splash with her duck genitalia, she made it her mission to recenter the female and convince other scientists that vaginas and their accoutrements were just as interesting and worthy of study as penises. "That's what my whole career is built upon," she says. That meant expanding her scope beyond birds—one of the flashier examples of genital coevolution—and tackling the rest of the animal kingdom. Recently, she delved into the mysteries of two-pronged snake vaginas, which parallel the male's dual members. To understand vaginal shape, she injects dental silicone into the vagina, lets it harden, and uses a toothpick to pull out the resulting mold. "I make vagina lollipops, Rachel," she told me in her lab, hunched over a dead snake. "This is what my life has come to."*

In 2015, a chance pairing brought Brennan together with Dr. Dara Orbach, a Canadian PhD student who was researching the sexual anatomy of bottlenose dolphins. They met up at a biology conference where Brennan was presenting on the importance of genital coevolution as part of a symposium on penis diversity. Meanwhile, Orbach was investigating a strange feature found in the reproductive tracts of dolphins, whales, and porpoises: a series of inner fleshy lids, like a stack of funnels, leading up to the cervix. These were known as "vaginal folds," and for decades, scientists had assumed that their main purpose was to

---

* Her inspiration for this technique comes from one of the very few researchers to have explored the shape and size of the human vagina: a retired anatomy teacher named Paula Pendergrass. In the 1990s, Pendergrass was asked by Tambrands, a tampon and diaper company, to look into the natural spectrum of vagina shapes for the purposes of optimizing tampon design. After having some very unusual conversations with her dentist, she realized she could use dental silicone to make vaginal molds, using a junior tampon as a "retriever." By making vaginal molds of 80 volunteer women, she found an impressive variety of shapes, which she categorized into five common ones: the conical, the parallel sides, the heart, the pumpkin seed, and the unfortunately named "slug-shaped." Although shape didn't turn out to matter much for tampons (given that they expand to fit their surroundings), it's possible that such information could be useful for tailoring vaginal devices and medications to specific populations, and for having a reference for vaginal reconstruction surgery.

keep sperm-killing seawater out of the uterus. But that function alone didn't explain the variability she was finding.

One year later, Orbach brought her frozen vaginas down to Brennan's lab, and the pair went about collecting more whales, dolphins, and porpoises that had died of natural causes. They set out to answer the question: what does a sea mammal's vagina look like in three dimensions, and how does it interact with a penis?[*]

To find out, they needed help from Dr. Diane Kelly, a genital-focused zoologist at University of Massachusetts who has spent decades finding creative ways to inflate disembodied penises and see how they fit into disattached vaginas. She came up with an ingenious strategy: to get the amount of pressure they needed, they used a mini beer keg to pump saline into the penises, inflating them so that they fit into thawed vaginas. Then they scanned both halves to make topographical maps of the genitalia interacting, and filled the vaginas with silicone to make three-dimensional casts. The result, for dolphins, looked something like a large, ridged dildo with a corkscrew-shaped addition at the tip. "It's kind of like Frankenstein science," Brennan says.

In some dolphins the vagina spiraled, not unlike a duck's, and was full of vaginal folds. Dolphin penises, in turn, ended in a cartilage projection, like a finger, which seemed to have evolved to open the sphincter-like lids and reach up to the cervix. Perhaps, they speculated, by slightly repositioning her body, the female could influence whether the male's semen hit the exact spot it needed to reach the uterus. "She may not have control over whether they copulate with her, but she has control over whether they inseminate her," says Kelly.

In 2018, the trio announced in a study that cetaceans displayed "unprecedented" vaginal diversity. As it turned out, it wasn't just

---

[*] Once, they cleared out the entire biology department while doing a particularly pungent dissection on a 40-ton whale. "You don't know the smell of anything bad until you have smelled a humpback whale vagina," says Brennan.

penises that could be used as species identifiers: Orbach found that these vaginas were so unique, she could often tell a species simply by examining its vagina.* It was the most diverse set of vaginas within any vertebrate group that we know of, says Orbach, now an assistant professor of marine biology at Texas A&M University.

As they dissected dolphin vaginas, Brennan couldn't help but notice something else. "We kept making these dissections looking at vaginas, and I'm looking at this enormous clitoris," she says. It made perfect sense to her that a dolphin would have a well-developed clitoris: they're known for mating year-round for reasons like pleasure and social bonding, and have been seen masturbating by rubbing against things like sand, other dolphins, and even eels. But when she went to the literature, she found that no one had described these massive organs—despite the fact that you'd almost have to go out of your way not to.

Brennan was familiar with the story of the human clitoris, and how it had long been neglected by science—she teaches about it in her class on evolution and human sexual behavior—so she knew what she was dealing with. Though no one could claim to having actually seen a dolphin orgasm, she could safely conclude that this clitoris was functional: It was dense with erectile tissue and blood vessels and shaped remarkably like a human clitoris. She and Orbach got to work writing yet another paper, this time describing dolphin sexual anatomy for the first time. "Where there is a vagina, there's usually a clitoris somewhere in the vicinity," she says. "It would seem like a wasted opportunity not to look at them."

---

* One of the few insects classified based on female, rather than male, genitalia is a leaf-dwelling praying mantis from Madagascar. It is named *Ilomantis ginsburgae* in honor of the late Justice Ruth Bader Ginsburg, a champion for gender equality.

Patty Brennan thinks of penis and vagina evolution as a dialogue between the sexes, a conversation in which each side eventually gets its say. But the conversation may not quite be an equal one. If you want to know which side is exerting the most pressure, it's probably the vagina. "The vagina was first. I will totally defend that," Brennan says.

To understand why, we have to go back to those restrictions that make the vagina less variable than the penis. Before penises existed, some female animals developed internal fertilization, and therefore required a tube to get eggs from the ovaries to the outside. And it is this original function, rather than the penis-receiving one, that may have shaped vagina evolution more profoundly—at least in humans, says Holly Dunsworth, an anthropologist who studies the evolution of reproduction at the University of Rhode Island.

In 2014, Dunsworth noticed an onslaught of articles in the popular media about the how the supposedly "huge human male penis" came to be. The most popular theory utilized Darwin's idea of female choice: women prefer larger penises, finding them more attractive and more pleasurable. Therefore they choose men with larger penises, leading men to develop larger and larger ones, and vaginas to play catch-up. Although the evidence was slight, it was a theory that clearly had cultural appeal: everyone knew that women loved big penises.*

"And I really was sick of reading all that," Dunsworth says. "I wanted to rebel." She went to the literature. That's when she realized that, in their eagerness to enshrine the penis, most people had completely forgotten about the vagina's second function, baby ejection—which, as Brennan points out, "is a pretty darn important function right there."

---

* Actually, compared to other primates, human males have relatively wide, but not particularly long, penises.

The heads of human babies are some of the largest among all primates. To give birth to these increasingly big-brained, huge-headed
infants, we had to evolve wider pelvises and wider birthing canals.
At rest—say, with your legs crossed, sitting down—there's hardly any
canal at all; the opening of the vagina is around an inch in diameter,
and the walls touch. But during childbirth, a human vagina opens like
a pleated accordion, stretching more than 300 percent to accommodate
a baby's head that is on average 13.75 inches in circumference—kind
of the opposite of a python eating a pig. More amazingly, it returns to
its normal size within six to twelve weeks—and nobody knows quite
how. "The fact that you can get something the size of a cantaloupe out
of the vagina and then have it go to even close to being normal is just
miraculous," says Dr. John DeLancey, a gynecologist who specializes in
pelvic MRI and worked with Helen O'Connell on her clitoral imaging
studies. "It's one of the most amazing things about the human body
that's never been studied."*

Dunsworth suggests that this could be why our vaginas had to stretch
over time: not for intercourse but for childbirth. As a side effect, penises
evolved to fit inside them. "It's pretty straightforward," she wrote in a
2015 blog post "Why Is the Human Vagina So Big?": Penises are just
playing catch-up. This explanation may not be as sexy, but according to
Dunsworth, it's a lot more scientifically sound: "Wouldn't you explain
the size and shape of the key by the size and shape of the lock?" she
wrote. "If there's an exceptionally human story for the great big human
penis, that exceptional story originates not in a woman's orgasms, not

---

* Other changes that happen during pregnancy: More blood is siphoned to the pelvis, which can enlarge veins and darken the labia to a purplish-bluish hue. Hormones
released by the placenta cause the muscles of the pelvis to soften and stretch. As a
result, the basket of muscles that line the bony pelvis (the ones you tighten when you
do "Kegels") curve downward into a more bowl-like shape—think of going from a
shallow soup plate to a more rounded cereal bowl. The clitoris may also change shape
and enlarge.

in her pornographic thoughts or her lustful eyes but in her decidedly unsexy 'birth canal.'"*

Most "why" questions in evolution are notoriously impossible to answer, and Dunsworth acknowledges that hers is just another theory. "It does come down to who's telling the story," she says. Her point is that, for too long, the field of human evolution has elevated theories that reinforce our idea of human exceptionalism. In the big vagina case, the theory also conveniently supports the long-held but little-supported assumption that larger penises are more attractive and pleasurable for women. By relying on these intuitive answers, we blind ourselves to potentially truer, or more imaginative, possibilities.

⌣

Here's an even more radical possibility: What if we looked beyond reproduction altogether? After all, genitalia, contrary to Darwin's claim, do far more than just fit together mechanically. They signal, symbolize, and titillate—not just to a potential mate, but to other members of a group. In humans, dolphins, and beyond, sex serves richer and more complex purposes than solely the transfer of sperm from one party to another. It can be used to strengthen friendships and alliances, make gestures of dominance and submission, and as part of social negotiations like reconciliation and peacemaking, argues ecologist and evolutionary biologist Joan Roughgarden, author of the 2004 book *Evolution's Rainbow: Diversity, Gender, and Sexuality in Nature and People.*

These other uses of sex may be one reason that animal genitalia are so weird and wonderful beyond your standard vagina/penis combo. Consider the long, pendulous clitorises that dangle from female spider monkeys and are used to distribute scent; the notorious hyena

---

* In a recent peer-reviewed paper, Dunsworth also cited other potential explanations for the size of the human pelvis besides childbirth, including a prosaic one: the female pelvis simply has more organs in it.

clitoris, which is the same size as the male's penis and used to urinate, copulate, and give birth; and of course, the showstopping genitalia that Darwin highlighted in "certain monkeys"—including the
rainbow-hued genitals of vervets, drills, and mandrills, and the red
swellings of female macaques in estrus—that may connote social status
and help different troupes avoid conflict. These diverse examples of
"genital geometry" (Roughgarden's term) serve a multitude of purposes beyond reproduction.

"All our organs are multifunctional," she points out. "Why shouldn't
the genitals be as well?"

Same-sex behavior is widespread throughout the animal kingdom.
In female-dominated species like bonobos, same-sex matings are at
least as common as between-sex matings. Notably, female bonobos
have massive, cantaloupe-sized labial swellings and prominent clitorises
that can reach 2.5 inches when erect. Some primatologists have gone
so far as to suggest that the position of this remarkable clitoris—it's in
a frontal position, as in humans, and unlike in pigs and sheep, which
have clitorises inside their vaginas—might have developed to facilitate
same-sex genital rubbing.

"It does seem more logistically favorable, let's say, for the kinds of sex
they're having," says primatologist Amy Parish, a bonobo expert who
was the first to describe bonobo societies as matriarchal. Primatologist Frans de Waal, too, has mused that "the frontal orientation of the
bonobo vulva and clitoris strongly suggest that the female genitalia are
adapted for this position."

Roughgarden has coined this rare clitoral configuration the "Mark
of Sappho." And, given that bonobos, like chimps, are some of our
closest evolutionary cousins—they share 98.5 percent of our genes—
she wonders why more scientists haven't asked whether the same forces
could be at play in humans.

These are questions that the current framework of sexual selection,
with its simple assumptions about aggressive males and choosy females,
renders unaskable. Darwin took for granted that the basic unit of nature

was the female-male pairing, and that such pairings always led to reproduction. He ignored inconvenient examples or wrote them off as exceptions, then pigeonholed males and females into narrow gender roles. Therefore, the theory he came up with—coy females who pick among competing males—only explained a limited slice of sexual behavior. Most evolutionary biologists who followed in his footsteps similarly treated heterosexuality as the One True Sexuality, with all other configurations as aberrations—at best, befuddling exceptions or amusing curiosities; at worst, a form of deceit that animals employ to pass on their genes.

The effects of this pigeonholing go beyond biology. The dismissal of homosexuality in animals, and the treatment of such animals as freaks or exceptions, helps reify negative attitudes toward sexual minorities in humans—including lesbians, gay men, intersex people, and asexual people, to name a few. Like Freud, Darwin's theories are often misused today to promote myths about what human nature should and shouldn't be. Roughgarden, a transgender woman who transitioned a few years before writing her book, could see the damage wrought by these stereotypes more clearly than most. Sexual selection theory "denies me my place in nature, squeezes me into a stereotype I can't possibly live with—I've tried," she writes in *Evolution's Rainbow*.*

Like Dunsworth, Roughgarden points out that biology is about storytelling. And thus far, sexual selection biologists have been stuck telling the same old yarn. Focusing on a few dramatic cases of sexual conflict—the "battle of the sexes" approach—obscures the stunning variety of forces that shape genitals. It also sidelines the many species in which the sexes cooperate and negotiate, including monogamous seabirds like albatrosses and penguins. "Biology need not limit our potential. Nature offers a smorgasbord of possibilities for how to live," she

---

* Just before her book was published, a journalist wrote that "some scientists privately wonder if—whether she likes to admit it or not—Roughgarden's own experiences of social exclusion have biased her view of the natural world." By contrast, few would think to question the biases that shape a white, cisgender man like Darwin, writing about the same topic.

writes. Rather than chaste Victorian couples marching two by two up the ramp into Noah's neat and tidy ark, "the living world is made of rainbows within rainbows within rainbows, in an endless progression."

To tell these kinds of new evolutionary stories, Brennan believes, biologists need to shed some of the Darwinian squeamishness they have inherited. Once, for instance, she investigated how the vaginas of spiny dogfish sharks change during pregnancy (they become more asymmetrical, because one gestating pup generally sits lower in the uterus). She was dismayed to see that anatomy textbooks often ignored the dogfish vagina entirely, focusing instead on the shell glands and ovaries. She has found this again and again: detailed descriptions of the uterus and surrounding organs, but for the vagina—nothing. "I think people were just prudes," she says. "They literally didn't want to write a subtitle that said 'vagina.'"*

That prudery has real scientific consequences. When Brennan looked to the human literature for a baseline comparison of how vaginas change during pregnancy, she found nearly nothing. "I was pregnant twice," she says. "I'm sure my vagina changed dramatically. I could tell things were different. How were they different? I don't know. I wish I knew."

Part of the reason Brennan is able to do this work is because of who she is, and how she was raised. She grew up in Bogotá, Colombia, with four sisters and attended an all-girls Catholic high school. The experience made it obvious to her that women could do anything men could do. Even though her school was a highly religious environment, in her slice of Colombian culture, "there was not a lot of taboo or 'these are bad things,'" she says. "Colombians are comfortable with sex talk." Words like "penis"

* Meanwhile, on the first day of class, Brennan asks all her students to yell "Vagina! Vagina! Vagina!" and then "Clitoris! Clitoris! Clitoris!" to loosen them up.

and "vagina" weren't shrouded in silence. Sex was matter-of-fact, just part of life. "To me, it's just such a basic thing that we should really know. I'm so fascinated by it that I just can't shut up about it."*

The first time she realized that others might find her intent focus on animal genitals even a smidge inappropriate was when she was rejected for a teaching job at a university back in 2009. Later, she ran into the man who had gotten the job instead. She was, she admitted to him, a little disappointed. He asked her: "Do you think that the topic of your research may turn people off?"

She had never, in fact, had that thought. "Up until that second, my only perception of the research I was doing is everybody thought it was amazingly cool," she says. Now she realized that here she was—a young, bubbly Latina woman—waxing poetic to elderly white men about duck penises without blinking an eye. "Yes, of course I can see a seventy-year-old dude sitting there being like, Oh my gosh, is this what my departmental meetings are going to be like from now on?" she says.

Yet other leaders in the field value the work she is doing. Eberhard, who is retired in Costa Rica, admits that while some groups of animals with well-studied genitalia do have more static female genitals, there are far more that simply haven't been studied. These are the ones that Brennan is illuminating. "I have looked at many, many, many vaginas in my career, and I have never seen one that I was like, Oh yeah, this is exactly what I thought it was going to be," she says. "I have always found something surprising. Or something that nobody knows."

When people introduce Brennan as a speaker at conferences today, they often lead with her work on duck penises—by now everyone knows about those. Brennan gently corrects them: she studies penises *and* vaginas. It doesn't help to only look at one or the other. Only by zooming out can we see them in their full range of variation and possibility.

---

* Her kids, she notes, often win the contest of who has the weirdest dinner conversations at home—not much beats dolphin vaginas.

In practice, that means erring on the side of vagina research, at least for now. "You can't just do penises, and you can't do just vaginas. You have to do them both," she says. "But here's the catch: We know so much less about vaginas than we do about penises that it's going to take a while to catch up to the basic vagina biology. And so it's going to have to be a lot of vagina for a while, before we can put more of it together."

A lot of vagina, indeed.

We don't actually understand what makes a healthy vagina at all.

—DR. CAROLINE MITCHELL

CHAPTER 4

Protection

(VAGINAL MICROBIOME)

D r. Ahinoam Lev-Sagie was at a loss. It was 2014, and yet another patient was in her office in tears of frustration. Lev-Sagie, a gynecologist who runs a clinic for hard-to-treat vaginal problems at Hadassah University Medical Center in Jerusalem, has seen everything—painful intercourse, postpartum infections, allergies to the body's own hormones—but this particular ailment was by far the most common and devastating. "I have patients that tell me that they don't date," she says. "Some of my patients were telling me they didn't have sexual intercourse for one year, two years, because they couldn't stand the idea that they smelled so badly."

One was a woman I'll call Alma. The forty-eight-year-old Israeli woman had always considered herself in tune with her body, and particularly her vagina, which she refers to as her yoni. ("I know her very well," she says.) Yet for the past three years, she had been baffled by a stubborn bacterial infection. She'd tried antibiotics. She'd tried probiotics. She changed her diet. Nothing worked. For most of her life, Alma had been confident about her body and her sexuality. Now she felt that both were out of her control. "I felt disgusting," she said. "You're not able to be clean enough for yourself. Whatever you do, you're not able to be clean." By the time she came to Lev-Sagie, she was on her third recurrence.

Lev-Sagie told Alma that she had a condition called bacterial vaginosis, or BV, an overgrowth of certain bacteria that naturally live in the vagina.* It affects nearly 1 in 3 women before menopause, adding up to 21 million women in the United States, according to the Centers for Disease Control. BV commonly makes itself known as a thin white or gray discharge, itchiness around the vulva, and a "fishy" smell caused by the bacteria-made chemicals cadaverine and putrescine. Yet despite how common it is, Lev-Sagie had no good treatments to offer. The best she could do was prescribe more antibiotics, which wouldn't stop a stubborn infection from coming back and could open the door to more infections.

Even as the words came out of her mouth, she realized that medicine could do better. She spoke with a colleague in her lab, a placenta researcher who had just read a groundbreaking new paper on the promise of microbial transplants to cure disease. That colleague got her thinking about poop.

Poop—or, more formally, fecal transplants—got its shining moment in 2013. That's when a group of researchers from Johns Hopkins announced a radical treatment for an infection by the gut bacterium *Clostridium difficile,* or *C. diff. C. diff* is known for its tendency to overgrow and take over the intestinal ecosystem following antibiotic use. It can cause severe diarrhea, inflammation of the colon, and even death in older or immune-compromised patients. Yet by infusing a patient's intestines with diluted fecal matter from a healthy donor—a process now performed via a colonoscopy—researchers showed that they could revitalize the gut milieu and dramatically restore a patient's health.

---

* BV, by the way, is the infection I described in the introduction.

Their trial was so successful that, halfway through, they switched all participants to the fecal transplant route.

Bacterial vaginosis, Lev-Sagie realized, was also an imbalance in the body's ecosystem. On a microscopic level, the vaginal ecosystem is very different from the gut. A healthy gut is crawling with around three hundred to five hundred different bacterial species, while the vagina is usually dominated by one main genus of bacteria, *Lactobacillus*. But the concept, Lev-Sagie believed, was similar: What if she could alter the vaginal environment by introducing new, healthy microbes into it?

There were already vaginal probiotics on the market, laced with lactobacillus, that promised to improve vaginal health. But they lacked scientific support and were largely known to be useless. (In the United States, probiotics generally fall under the category of supplements, which aren't regulated by the FDA.) Lev-Sagie had a more ambitious idea: to transplant an entire vaginal microbiome from a healthy woman into her patients with persistent BV. Like a fecal transplant, the idea was to reseed the vagina with bacteria that would help a woman grow back her natural balance of protection. By transforming her vaginal ecosystem, Lev-Sagie hoped to break the vicious cycle of infection and reinfection.

In some ways, however, vaginal transplants aren't quite the same as fecal transplants. *C. diff* is widely recognized as a serious, life-threatening problem. Nearly half a million Americans get it yearly, and an estimated 15,000 die from it. Extreme diseases, as Hippocrates is said to have written, call for extreme cures: once a treatment of last resort, fecal transplants are now a standard route of treatment for persistent *C. diff*, and are being studied for their potential to treat ulcerative colitis, irritable bowel syndrome, and other chronic gut issues.

Vaginal transplants, meanwhile, have an extra strike against them.

Besides the "ick" factor, there's the fact that "women's issues" are largely underappreciated by medicine. "BV is a quality-of-life-threatening infection," says Dr. Caroline Mitchell, a gynecologist who runs a vulvovaginitis lab at Massachusetts General Hospital and studies the interactions between humans and their microbes. "But very few people seem to care much about women's sexual health and quality of life." Like Lev-Sagie, Mitchell has seen firsthand how devastating recurrent BV can be on women's self-esteem, relationships, and health. Some of her patients spend hundreds of dollars on unregulated vaginal probiotics. Others turn to dangerous, unproven remedies like steam douching.

These kinds of details aren't enough to convince organizations like the NIH to fund this work, says Mitchell. That's why, when she makes her argument in grant proposals, it's never just about women's suffering. It's about broader public-health outcomes like HIV, cervical cancer, and preterm birth. Women with BV have double the normal risk for preterm births and miscarriages, and a much higher risk of contracting HIV and other STDs if exposed. "That's how you sell it as being important," she says. "Women's quality of life, women's sexual health, women's symptoms are not usually compelling enough. Just 'vagina problems.'"*

In fall 2021, Mitchell was starting a vaginal transplant trial at Massachusetts General Hospital with the goal of sifting out what elements make up a healthy bacterial environment. In May, I spoke to one hopeful participant, twenty-four-year-old Victoria Field, who had applied to join the trial. Field told me she turned off her GPS location on her phone when she drove from her home in Ithaca to Boston because she didn't want anyone to know she was in a clinical trial for a vagina problem. "It's very personal," she says. "But it's like a double-edged sword, because it's so personal and embarrassing that I don't want to

---

* Meanwhile, millions of dollars and numerous clinical trials a year are devoted to overcoming erectile dysfunction, which also never killed anyone.

talk about it. But by not talking about it, I'm also contributing to the problem of nobody wanting to talk about it." In the end, she chose to share her experience with me to help lift that stigma.

Lev-Sagie ran into these biases as well. When she first floated the idea of a trial, male colleagues told her this wasn't worth pursuing. "You're not going to die from having a fishy odor," they said. But the reaction she got from patients like Alma confirmed for her that there was a deep need for this—deep enough to overpower the initial "ick" reaction. After she set up a trial on clinicaltrials.gov, she was bombarded by requests from women as far away as Europe and the United States. It was the second-most popular trial on the site, after a study focusing on amyotrophic lateral sclerosis (AMS), a fatal motor neuron disease. "Just imagine that people are willing to come to Israel in order to participate," she says. "It gives you some sort of understanding of how bad BV can be."

She started the pilot study in 2015, with Alma as one of five participants. First, all participants were treated with vaginal antibiotics. A week later, they were given the transplant: a glass syringe dipped in the vaginal fluid of a healthy woman, swiveled in her vagina for a few minutes. Alma had two transplants in the first week, and after feeling symptoms again a few months later, she got one more. When Lev-Sagie looked at her vaginal smear under the microscope, it was clear her microbiome had been transformed. Thanks to an intimate gift from a person she would never meet, "I got my life again, I got my freedom again," says Alma. "No more guilty, no more dirty, no more shame." When I spoke to her two years after the trial, she was still symptom-free.

Today, Lev-Sagie thinks about vaginal microbiome transfers differently. Studies show that long-term female sex partners end up having remarkably similar vaginal microbiomes to each other. Perhaps getting a vaginal microbiome transfer is more like having a new sexual partner, she thought. Except in this case, your new partner has had all the possible examinations and tests you could think of—and promises to change your vaginal ecosystem in a *good* way. More research is needed, Lev-Sagie acknowledges. "We will have to show that it works in a randomized controlled trial, but I was convinced that it's really working."

Vaginal transplants were a logical solution with a proven foundation—so logical that at times, both Lev-Sagie and Mitchell wondered why no one had done this before. Why had it taken so long for this message to get through to women: that BV isn't some shameful disease but an ecosystem shift in the vagina that can be treated just like one in the gut? Well, it turns out the history of researching vaginal infections isn't exactly pretty. Doctors of the past did try to do this kind of transplant experiment, but in the reverse: they infected healthy women with bacterial vaginosis, just to prove they could.

———

In 1955, Dr. Herman L. Gardner, a bacteriologist at Baylor University in Houston, Texas, zeroed in on a tiny, round bacteria that tended to congregate in some women's genital tracts. He pointed to it as the prime culprit behind a condition then called bacterial vaginitis, and today known as BV, whose most obvious symptoms were itching and burning of the vulva, and a gray discharge with a "disagreeable odor." "While the disease is not a serious one, it is physically and esthetically objectionable, and has undoubtedly contributed to the popular belief that all vaginas are all tainted and in need of frequent douching," Gardner wrote.[*]

To prove that this bacterium was the responsible party, he and a colleague ran an appalling experiment: They transferred vaginal fluid from fifteen women with BV into women who had never had the condition before. Some of the receiving women, who came from Gardner's "volunteer clinic," were pregnant during the trial. Within

---

[*] Note: They are neither. I can't believe I have to write that.

a week, the majority developed symptoms of BV, which the researchers allowed to go untreated for up to four months. When they looked
at their vaginal smears, the round bacteria now dominated the landscape, crowding out other cells by a ratio of 100 to 1. Many of the
women's husbands also had the bacterium in their urethras, showing
they had been infected. This was undeniable proof, he claimed, that
this bacterium was "the primary, if not the sole, etiologic agent of
the described disease."*

In 1980, near the end of his Gardner's life, the bacterium was
renamed *Gardnerella vaginalis* in his honor. His experimental subjects
would earn no such recognition.

While Gardner referred to these patients as volunteers, it is impossible to say whether they knew what they were agreeing to; he left that
part out of his paper. But his experiments are part of a larger medical tradition of relying on populations who were captive, vulnerable,
and unable to fully consent—a tradition that includes the infamous
Tuskegee Syphilis Study, early testing of the Pill on Puerto Rican
women who weren't fully informed of the risks, and the use of Henri-
etta Lacks's cervical cells without permission to create an immortal cell
line for research. In fact, that is the origin story of the modern field of
gynecology. It goes back to James Marion Sims, the Southern slave-
holder and doctor who would become known as the "Father of Modern Gynecology."

By many accounts, American gynecology began in a makeshift hospital Sims built on the corner of his family farm, ten miles from Mont-
gomery, Alabama. Although he disdained the specialty—"If there was
anything I hated, it was investigating the organs of the female pel-

---

* Although Gardner referred to BV as an STI, that is still under debate. The condition is rare in women who haven't had sexual contact, and can be transferred between
sex partners of all genders (in men, BV-associated communities can establish inside
the urethra and foreskin, which also have their own microbiomes). But really, it's
more of a disturbance in the ecosystem rather than a disease caused by one microbe.

vis," he would write in his autobiography—necessity drove him to the field. Other slaveholders came to him because their enslaved women were suffering vesicovaginal fistulas, openings between the walls of the bladder and vagina caused by damage from traumatic or obstructed labor. These tears caused urinary incontinence and made them "unfit" to work. "The engine of slavery rests on enslaved women's healthy wombs and healthy births," says Deirdre Cooper Owens, a historian of race and medicine at the University of Nebraska–Lincoln and author of *Medical Bondage: Race, Gender, and the Origins of American Gynecology.* It was therefore in the interests of both slaveholders and doctors to protect these women's reproductive health.

Sims greased that engine. He sought out women who suffered from this malady and performed experimental surgeries on them in an attempt to develop a cure. "There was never a time that I could not, at any day, have had a subject for operation," he would write. Of the dozen or so women he wrote about in his medical notes, we only know the names of three: Lucy, Betsy, and Anarcha. In 1849, after four years of experimentation—including more than thirty experimental surgeries on Anarcha alone, who was seventeen and had recently given birth when her time with Sims began—Sims finally "perfected" his method of repairing fistulas, by using silver sutures instead of silk and draining the bladder after the operation with a catheter or sponge. His technique took off around the world, and is still widely used today.

Medical journals of the time emphasized the difference and otherness of Black women—including false stereotypes about elongated labia, hypersexuality, and the ability to withstand great pain. These stereotypes persisted despite observations that clearly belied them. Sims himself noted that "Lucy's agony was extreme" and that medical assistants had to restrain her during his surgeries. Once, he left a sponge inside her that allowed urine to seep into her vagina. Sims later admitted that these procedures were "so painful, that none but a woman could have borne them." He claimed that his patients willingly agreed to them, but once again, we only have his word to go by. And since these women were considered human property, they did not have the ability to truly consent.

Sims considered himself a lone medical pioneer. Yet his advancements were made possible thanks to the larger system of American slavery. Under it, Black women were not considered human enough to be free, nor human enough to be spared great pain. Yet they were human enough that Sims could apply his findings to white women. It was a form of convenient cognitive dissonance: Doctors extracted medical knowledge from the bodies of women, just as slaveholders extracted labor, in the name of science and exploration. Recalling the day he developed his signature medical tool, the vaginal speculum, using the handle of a pewter spoon, he wrote: "Introducing the bent handle of the spoon I saw everything, as no man had ever saw before. The fistula was as plain as the nose on a man's face." The speculum, his biographer would write, was to diseases of the womb as the telescope was to astronomy: science's triumph over nature.*

Sims would become one of the foremost gynecological surgeons in the country. In 1855 he founded New York's Women's Hospital, a charity institution that treated thousands of poor and immigrant women—while also relying on those same patients as research subjects. He later helped found the American Gynecological Society and served as president of the American Medical Association. For nearly a century he was memorialized by a bronze statue on a granite pedestal in New York's Central Park. "His brilliant achievement carried the fame of American surgery throughout the entire world," read the plaque, "in recognition of his services in the cause for science and mankind." The city finally removed it in 2018, following a review of "symbols of hate" in the wake of white supremacist protests in Charlottesville.†

---

* To examine his patients, he would later have them lay on their left side with one arm behind their back and one knee higher than the other—what would later become known in medicine as the "Sims position."

† Today Sims is often cast as a "manically evil man," a figure akin to Nazi doctor

Meanwhile, the women he experimented on sank into obscurity. Their voices would be lost to history. Or would they?

When Cooper Owens first came across Sims in lectures and twentieth-century medical texts, she wondered about his patients. The history of medicine was filled with "fathers"—the father of the C-section, the father of endocrinology, the father of ovariotomy (the surgical removal of diseased ovaries)—but, ironically, there were no mothers. She set out to recover the voices of those who had been left out: the enslaved women who bore his experiments and made his work possible. She scoured plantation owners' ledgers, census records, and Sims's medical writings for any mention of Betsy, Anarcha, and Lucy.

Piecing together these fragments, she realized that these women were more than just the bodies on which Sims built the foundations of modern American gynecology. They were active participants in that enterprise.

Two years after he had established his temporary hospital, Sims had not cured anyone. His two white male medical apprentices, who had been serving as surgical assistants, quit. Sims elected to have his patients take over. Lucy, Betsy, and Anarcha learned to restrain patients during surgeries and clean and dress surgical wounds. "Nobody had ever thought of them as nurses, as surgical assistants," says Cooper Owens. "But they're doing literally the same work. The white men who worked with him before, they aren't doing anything different. They're restraining patients. They're watching his technique. They are apprenticing, without it being called that."

There are no records of what happened to these women after their time with Sims. However, Cooper Owens found ledger records of other enslaved patients who later became nurses and midwives, leading her to believe the same might have happened for them. Whether they were

---

Josef Mengele, says Cooper Owens. Yet for her, it is Sims's ordinariness, rather than his uniqueness, that makes him so important to understand in the history of gynecology: "He is representative of a system that was already set in place."

able to continue using their skills or not, it was likely that they knew more about the repair of obstetrical fistulae than the most well-trained American doctors of the era. "When we think about the advancements of gynecology and obstetrics, we cannot separate it from slavery and from enslaved women," she says. For this work, she suggested that Lucy, Betsy, and Anarcha be named the Mothers of Gynecology.

Though the women Gardner experimented on were not enslaved, they also likely never had the opportunity to consent to such a procedure. Given this kind of legacy, you can imagine that modern researchers would want to tread carefully when it comes to experimenting with moving stuff from one vagina to another. "The creepy history, I think, has made it a little bit fraught," says Mitchell. Nevertheless, she believes the time for vaginal transplants has come, because there's so much to learn about this little-known ecosystem—and so much to gain in the realm of women's health. "That's my pitch to grant funders," she says. "No matter what happens, we will move the field forward."

It's time for a new vision of the vaginal microbiome—one grounded not in shame and fear but in wonder and diversity.

⁓

Your vagina is another planet. If you could shrink down to the size of a grain of sand and go between your own legs, you'd find a wondrous realm of humid jungles, cool caves, and viscous pits of mucus created by your teeming ecosystem of microscopic life. Like your gut or your mouth, your reproductive tract is home to billions of microbes, which work together to repel disease and create the ideal conditions for you. Its landscapes are populated by clusters of long, thin rods and hordes of tiny round balls that cling to its contours. These microbes live together in a delicate balance, spewing acid to stop would-be colonizers from worlds far-off (tampons, toys, penises) or nearby (the anus).

In the past decade, genetic-sequencing technologies and a growing appreciation for the microbial world have revealed that none of us is an island. We all live with 39 trillion or so of our closest friends—bacteria, viruses, and other micro-organisms that make their homes in our mouths, our intestines, even our brains. The human body harbors at least as many microbial cells as it does human cells: "There are more bacteria in your gut than there are stars in our galaxy," writes science writer Ed Yong in his 2016 book *I Contain Multitudes.* This new way of looking at the world—as one not of individuals but of invisible partnerships—has left no part of the body untouched.

If you think of each microbiome on your body as an ecosystem—a desert, a forest, an open tundra—the vagina would be a humid wetland, full of warmth, moisture, and peril. Just one liquid milliliter of vaginal secretion contains up to a billion individual bacteria, which fall into three hundred or so different species. Often it is one group that dominates: Lactobacilli, long hailed as the "good guys" of the vagina. These are the keystone species of the vagina: Like redwood trees, their presence sculpts the environmental niche that allows others to thrive. From the same genus of bacteria found in cheese and yogurt, these unassuming cylinders ferment sugars into lactic acid, keeping the pH of the vagina about as low as a glass of red wine (pH 3.5 to 4.5). That biting acidity helps keep unwanted bacteria at bay.

This is crucial, because the vaginal microbiome is your first line of defense against all manner of threats. Think of it as an extension of the immune system, one of the body's barriers between you and not-you. Taken together, these bacterial communities can be thought of as a unified barrier. But Mitchell and others see them as less like a standing army and more like a richly populated garden, with new species coming into bloom and others fading away. Vaginas change day by day, hour by hour. This dynamism makes for a highly effective defense: when disrupted, the vagina tends to mobilize and bounce back to some kind of equilibrium. When Simone de Beauvoir wrote that "the body is not a thing, it is a situation," she may as well have been talking about the vaginal microbiome.*

But at times that barrier can get catastrophically disrupted, leaving it vulnerable to unwanted shifts in its tenants. Having a few weeds in the garden is fine; it's when they overgrow and trample out the rest of the flora that it becomes a problem. The list of things that can potentially disrupt the vaginal ecosystem include: lubricants, antibiotics, IUDs, hormonal surges, douches, menstrual blood, ejaculate. (Semen, funnily enough, has a higher pH that can lower the vagina's acidity.) All of these events can kill off lactobacilli and encourage bacteria like *Gardnerella*, which are already present in the vagina in lower numbers, to proliferate.

These are some of the things we know. But thanks to those factors mentioned earlier—the "ick" factor, the dark history, the fact that no one cares about women's health—there are far more that we don't. For instance: Whether there is a similar but unique microbiome in places like the upper reproductive tract, Fallopian tubes, and uterine lining, a question relevant for fertility studies and procedures like IVF. Given that nearly every crevice of the body has its own microbiome, it's likely— yet the answer remains murky and controversial.†

Even when it comes to the superheroes of the vagina, lactobacilli, we still don't know exactly how they create the protective effect that they do. Nor do we know why some women are able to fight off infection easily, while others succumb again and again. "We know what is associated with good reproductive health outcomes," says Mitchell. "We just don't know how to get there. And so the question is, does it matter the seed, does it matter the soil, does it matter the fertilizer—we don't know." Trials like hers and Lev-Sagie's are some of the first bids to find out.

---

* Though, to be clear, she was not.

† Sampling these regions is challenging, but there may be another factor at play: the assumption that female reproductive organs are supposed to be pure and sterile.

Imagine looking through a microscope into two Petri dishes. The first contains a culture taken from a lacto-dominated vagina, while the other is from a woman with BV. In the first, your eye will immediately be drawn to couple of pink, jelly-like blobs with red centers. Those are epithelial cells, which form the lining of the vagina as well as other surfaces of the body—skin, blood vessels, organs. In the vagina this lining is thick and plump, shedding three to four layers of cells a day. This is what makes up typical vaginal secretion: bacteria, cells of the vaginal lining, and mucus. Around the pink blobs are clusters of rod-shaped cells with rounded ends, stained purple. These are the lactobacilli, sworn defenders of the vagina.

Peer closer, and you'll see that not all rods look alike. Some are longer and more noodle-like, while others are more compressed, like flattened soda cans. The long ones are *L. crispatus*, which "has been kind of crowned the king of lactobacillus, which I tend to agree and disagree with," says Dr. Jacques Ravel, a professor of microbiology and immunology at the University of Maryland. They spew out a kind of lactic acid that is particularly lethal to invading bacteria. The more compressed disc shapes, by contrast, are *L. iners*. They produce a different type of lactic acid that seems to be less effective at deterring pathogens and more resistant to antibiotics. Species like *L. iners* may still assist the vagina in some way. "We just don't know enough to say that any of them are not good, or better, than the others," Ravel says.

It gets more complicated. Within a single species, like *L. crispatus*, there can be thirteen strains in the same woman. These strains likely work together, each performing slightly different functions. Though the vaginal microbiome looks from afar like a dictatorship, it's more of a negotiation, with thousands of minute interactions and partnerships happening under the surface.

In the second Petri dish, there are no rods in sight. Instead, the pink blobs are positively covered in smaller, rounder bacteria, many of them clinging like ants to abandoned cookies. Others float around in space in grape-like clusters. "It's like night and day," says Ravel. Most of these are *Gardnerella*, but some have other names, like *Prevotella bivia*. What unites them is the shared quality of thriving in a low oxygen, high-pH environment. Because they are associated with a greater risk of infection, these oxygen-hating bacteria have traditionally been dubbed the bad guys. Whenever they get a chance, they usurp lactobacillus and take over the kingdom, knocking the vagina's balance off kilter.*

There are myriad reasons why women end up with this vaginal milieu. According to Ravel, a major one is vaginal antibiotics. For women with BV, chlamydia, gonorrhea, or syphilis, they are often the only option. But these indiscriminate killers wipe out everything in their path, including protective lactobacilli. Their scorched-earth policy creates a window for things like yeast, already present in the vagina in low numbers, to overgrow like a weed. (Boric acid, the last-resort treatment I was given, can have an equally dramatic effect.) Studies have found that antibiotics increase a woman's chances of being colonized by the fungus, and 1 in 5 women will develop a yeast infection after an antibiotic regimen. Even when the lacto does grow back, it's usually in the less-protective form of *L. iners*.

There's a reason we've relied on antibiotics for almost one hundred years, using them to vanquish everything from strep throat to UTIs: they're highly effective at stamping out the original bacterial infection. But now we may be paying the price. Ravel believes that the overuse of antibiotics to cure STIs has become a major public health issue, leading

---

* The "healthy" smear may also have a smattering of these smaller bacteria as well, particularly if the woman has fought off an infection before. Like scars, they remain as marks of past trauma.

to weakened protection and setting up women for reinfection by the original pathogen.

———

Another harmful practice is douching: cleansing the inside of the vagina with water, vinegar, or fragrance. The douching myth goes back to at least the nineteenth century, when products like the best-selling Lydia E. Pinkham's Sanative Wash were injected into the vagina via syringe to treat unpleasant odors or unusual discharge. "The Sanative Wash is useful not only in cases of grave ailments, but is of great value to check slight disturbances of the secretions," the bottle label claimed. "As a deodorant, it prevents embarrassment." It was certainly more appealing than other standard medical cures of the time: applying nitrate of silver to the vaginal mucus membrane, or lotions of lead and mercury to the labia.

In the 1920s, the home-cleaning brand Lysol followed suit, marketing itself as a "safe and gentle" feminine cleanser. Just as it disinfects surfaces and kills germs in the home, Lysol promised to remove unwanted bacteria from the vagina. This was also code for selling Lysol as a contraceptive in the days before birth control was legal—if it killed germs, the thinking went, it would also kill sperm.* Ads promised that Lysol-douching would "keep you desirable!" and safeguard your "dainty feminine allure." In reality, it often led to inflammation, burning, and even death. Although Lysol is rarely used today, nearly 1 in 5 American women still douche. Unfortunately, these attempts to "cleanse" the vaginal flora strip the vagina of its natural protection and irritate the vaginal mucosa, which can lead to worse odor, more discharge, and new infections.

Other practices can have equally dramatic effects. In South Africa, researchers trying to slow the spread of HIV have found that teenage girls get the virus at a rate five times that of their male peers. Science

---

* No, it didn't really work.

writer Olga Khazan explains that the reason has to do with a system of sugar daddies—called "blessers"—who target impoverished girls and women for sex in exchange for gifts and financial support. These men often demand sex without condoms, and are thought to prefer a vagina that is "tight and dry," Khazan writes for *The Atlantic*. To achieve this, "it's common for women to coax their nether regions into this state by stuffing them with various powders, ashes, and even chewing tobacco." Once again, this weakens the vagina's defenses.

Like Mitchell and Lev-Sagie, Ravel is at work on a solution to these vaginal insults. But instead of transferring one woman's microbiome to another, he's trying to reverse-engineer the formula for a healthy vagina in the lab. LUCA Biologics (an acronym for Last Universal Common Ancestor), the company he founded in 2019, uses the library of vaginal data he has collected over the past fifteen years to create "live biotherapeutics"—supplements that include lactobacillus but also nutrients and molecules that form a better environment for lacto to grow. Studies show that vaginal probiotics containing only lactobacillus rarely colonize the vagina for long. "They don't have the proper nutrients, they don't have the right sugar, they don't have the right pH," says Ravel.

Others caution that, before we are in a position to fix anything, we need to do more basic research on the natural inhabitants of the vaginal microbiome. Our lack of knowledge, says Mitchell, is profound: when she first started studying the impact of the vaginal microbiome on genital shedding of HIV, she realized that "we don't actually understand what makes a healthy vagina at all."

———

In 2010, Ravel and his colleagues set out to define what makes up a healthy vagina. Using next-generation genetic sequencing, they analyzed the vaginal microbiomes of nearly four hundred asymptomatic North American women to see what bacterial communities could be considered "normal." The researchers asked the women to self-identify as white,

Black, Hispanic, and Asian, and found that their microbiomes generally fell into five community types, with four of them dominated by lactobacillus. The fifth represented a sort of détente between diverse groups of bacteria—including *Prevotella* and *Gardnerella*—with few lacto rods.

This state was more common among Black and Latina women. In white women, more than 40 percent had *L. crispatus* as their dominant microbe, while 20 percent had *L. iners*. In African American women, 40 percent had *L. iners*, and 40 percent had few or no lactobacilli whatsoever. Did this mean many Black and Hispanic women had "unhealthy" vaginas? Hardly. "If accepted at face value, this common wisdom suggests that although most Asian and white women are 'healthy,' a significant proportion of asymptomatic Hispanic and Black women are 'unhealthy'—a notion that seems implausible," Ravel and his colleagues wrote. More likely, they said, it was that researchers hadn't yet characterized what a healthy microbiome might look like in these women.

It all came down to context: What causes issues in one woman may be perfectly healthy in another. It could be that what they had characterized as the "bad guys"—*Prevotella* and *Gardnerella*—did no harm in a different environment. These women might even have developed slightly different bacterial partnerships that accomplished the same ends. In some cases, Ravel acknowledges, problems could arise: if they have multiple new partners, or don't remember to protect themselves during intercourse, they may have a higher likelihood of getting an infection. "And that's a bit of a problem," Ravel says. "It doesn't mean it's bad, it just means that it's maybe not optimal." But overall, he concluded that diverse microbiomes "are common and appear normal in Black and Hispanic women." Like eating a Reese's Cup, there's no wrong way to have a vagina.

Thinking of diverse microbiomes as potentially healthy—and using terms like "not optimal" instead of "unhealthy"—can be seen as progress, says Dr. Jessica Wells, a women's health researcher at Emory University's School of Nursing who studies cancer and HIV prevention in at-risk populations. Previously, microbiome researchers considered high numbers of lactobacillus and a low pH to be synonymous with health. Uncoincidentally, these were the ecosystems associated with white women, who were most often the subject of vaginal microbiome studies. As a result, we've centered the idea of a "healthy" vagina on white women, says Wells. Most of the solutions we have for BV—antibiotics, and for persistent infections, boric acid—have been shaped by the idea that white women's vaginas are the healthy ones.

"Medicine is supposed to be unbiased," she says. "But when we look at the bigger picture of it, we can see that it is far from that."

When Wells first started reading papers that compared vaginal microbiomes on the basis of race, she noticed that many treated white women as the standard, and anything that deviated as unhealthy or maladapted. More disturbingly, they seemed to take for granted that the different health outcomes between racial groups were due to biological, not social, differences. This assumption made it impossible to untangle whether these differences ultimately stemmed from causes that weren't rooted in race or genetics—for instance, the fact that a woman of color in the United States lives with higher levels of stress and the hormone cortisol her whole life.

Black women living in the United States have the highest rates of infertility, preterm birth, infant mortality, maternal death, and STIs like gonorrhea. Many of these health disparities are directly related to social factors: inequities in income, housing, education, access to medical treatment, and racism. A 2006 study suggested that chronic stressors can hijack the immune response and make people more susceptible to infections like BV. One mechanism may be increased inflammation, which can increase risks for pregnancy complications and cardiovascular disease.

"For me, living as a Black woman in the US, unfortunately race is always on my radar," Wells says. "I'm maybe more aware and sensitive to blind spots that other scientists have no clue to."

Studies have found that Black women have more than double the risk of BV than white women do.* But the crucial question is why: "What is it about living as a person of color in the US that makes you at higher risk for BV?" says Mitchell. "That is a profound question that we haven't answered."

For Wells, the answer most likely lies not in genetics but in a long history of segregation, discrimination, and unethical experimentation that has led to a distrust of many aspects of mainstream medicine. In 2020, she published one of the few systematic reviews on Black women's microbiomes that pointed to potential social and environmental factors at play. "There is a budding narrative that women who are nonlactobacilli-dominant have a vaginal microbiome composition that is dysbiotic and unhealthy," she wrote. "However, research has shown that the microbiome is influenced by a host of factors, and a deeper understanding of the role that race and ethnicity plays in the composition of the vaginal microbiome is still warranted."

"We're all women," she says now. "We all come with different cultural backgrounds, different environments, different stressors. And perhaps our vaginal microbiomes are a reflection of our social factors."

———

Over the years, Ravel has looked at thousands of women's vaginal microbiomes under the microscope. The vast majority are dominated by one of the four common species of lactobacillus, or no lactobacillus. But in a few—less than half a percent—he's noticed something odd. These women had a different dominant genus of bacteria altogether: a branched, Y-shaped rod known as *Bifidobacterium*. It was

---

* Although even that data is conflicting, Wells has found. A BV diagnosis still requires a doctor to interpret a vaginal smear, and Black women could be being overdiagnosed due to implicit bias.

usually found in the intestinal tract, but somehow, it seemed to have gotten into the vagina.

A common folk remedy for itchy, irritating vaginal ailments like yeast infections is to dip a tampon in yogurt and insert it into the vagina. (From the classic women's health handbook *Our Bodies, Ourselves*: "Some women have had success with . . . acidifying the system by . . . inserting plain, unsweetened, live-culture yogurt in the vagina.") If you look on the back of any yogurt carton, you'll find *Bifidobacterium* (aka Bifidus), a live culture often advertised as being beneficial for gut health. Ravel suspected these women had tried yogurt douching, and the strain had taken root.

That we could be colonized by bacteria from our food is one of the most intriguing theories for how we ended up with our unique vaginal milieu. Of all other animals in the world—baboons, rabbits, mice, rats—our lacto-rich vaginas stand alone. This is a mystery, because great apes, like us, have to deal with live birth and the threat of vaginal infection. So why didn't they evolve the same symbiosis we have with lactobacilli? "It's a big puzzle in human evolution," Ravel says.

He has a theory. Around ten thousand years ago, nomadic humans settled down, tilled the soil, and became agrarian farmers. They domesticated livestock and bred plants, shaping their environments and bodies in the process. One of the developments that happened around this time, all around the world, was fermentation. By combining sugars with microbes like yeast, humans were able to preserve valuable food for far longer, and make it far tastier: think cheese, pickles, Miso, kimchi, yogurt, and tempeh. The key to this process was lactobacillus. As humans happily munched their lacto-loaded foods, some of the, er, excreted remains, may have made their way to the vagina.

Those bacteria would have found themselves in the warmest, wettest, most welcoming environment they had ever encountered. Lacto remade the vagina in its image, acidic and inhospitable to invaders. That quality would have conferred a great benefit to infection-prone humans, so that the two would likely have developed a symbiosis. Over time, the bacteria could have evolved to become more specially adapted to the vaginal environment, until they diverged enough to look far different under the microscope than the ones found in cheese and yogurt. Such a beneficial partnership would likely have occurred many times in many cultures, with the result that there are several different strains of dominant lactobacilli.

The diet theory is compelling, but Ravel cautions against reading too much into it: "It's very hard to prove or disprove."

For the fraction of a percent of women who do show up with yogurt strains in their vagina, these strains are likely setting up shop on only a temporary basis, says Dr. Willa Huston, a microbiologist at the University of Technology–Sydney who studies the links between chlamydia, infertility, and the vaginal microbiome. "It might feel like a bit of a salve, the cream might relieve some of that itchiness," she says. "But I don't think it's going to fix your microbiome." Yet Huston understands the instinct to try something, anything, when mainstream medicine has failed to offer any solutions. "In a way it's kind of worrying, but I don't know that there's any evidence of harm either," she says. "And if it's giving them some relief, can you blame them?"

⸺

Researchers aren't yet turning to yogurt douching. But they are recognizing that traditional BV cures like antibiotics (and boric acid) don't work in the long run. The idea of a protective microbiome transplant is promising, but for it to be a viable solution globally, it needs to be cheap, stresses Dr. Jo-Ann Passmore, an immunologist who studies genital-tract immunity and STIs in Cape Town, South Africa. Passmore's primary concern is HIV transmission rates, which are higher in her region than anywhere else in the world. And BV, which afflicts a large percentage of South African women, is a strong predictor of HIV transmission.

Passmore is working with Mitchell to expand her trials to South African women, testing slightly different variants of microbiome transplants tailored to that population. That means researchers will need to consider moving away from a prohibitively expensive personalized medicine model, and toward a mass-produced vehicle. In many parts of South Africa, Passmore points out, girls can't even afford pads or tampons. "I think a product is feasible and likely and it's going to make a huge impact," she says. "But whether and how it would be affordable where we live is another question."

An alternative delivery model she's looking into is fermented foods: the idea of enriching processed foods with forms of lactobacillus and other bacteria, the same way that cereals are enriched with nutrients like iron and vitamin B12. She's currently running trials with mahewu, a traditional South African breakfast drink made of fermented maize and sorghum. But she isn't ruling anything out; she says that she wants to see these therapeutics in the bathwater and water supply. "We're just trying to think creatively about what alternatives there would be," she says.

Defining effective vaginal probiotics would open up other possibilities for improving women's health, adds Lev-Sagie. A key shift in the vaginal microbiome comes at puberty, when a surge of estrogen plumps up the lining of the vagina with cells rich in glycogen, lacto's favorite food source. Vaginal transplants might be used at that time to ensure a girl has a complete transition to lactobacillus dominance and prevent disease later on. Or, probiotics could be given to women with less severe cases of BV, as an alternative to antibiotics. Stronger vaginal defenses could help protect women against numerous forms of STIs, including those that cause cervical cancer.

What's clear is that the vaginal microbiome is in need of new solutions and creative thinking. Jessica Wells, the women's health researcher at Emory, would like to see studies that spend less time sifting women's microbiomes into racial categories and more asking "So what?": how do we treat women, all women, not just those with lacto-dominant microbiomes? Whether it's living vaginal transplants, synthesized lactobacilli supplements, or probiotic-laced foods, the goal should be better solutions for everyone. "Putting boric acid in women's vaginas is just archaic," Wells says. "We can do better, scientists."

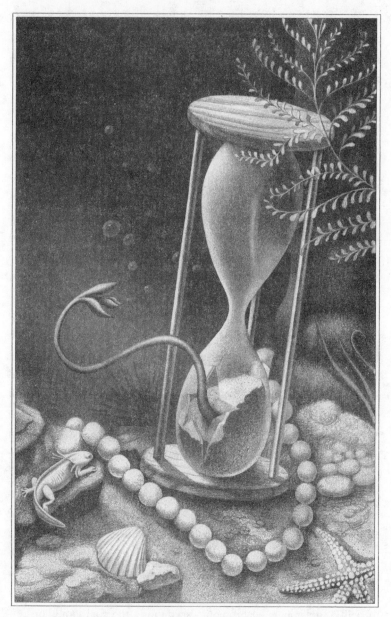

Even if I had Liz Taylor's diamonds and I looked at them, I wouldn't think
they were as glamorous as seeing this egg and the spermatozoa whirling.

—MIRIAM MENKIN

CHAPTER 5

Creation

(EGG CELL)

M iriam Menkin stands outside the operating room, tapping
her foot and holding a glass jar. It's eight a.m. on a Tuesday
in 1944, and she is anxious. Tuesdays are Dr. John Rock's
surgery days, when he spends the morning performing hysterectomies
and removing ovarian tumors. Rock is Miriam's boss and the director
of the fertility clinic at the Free Hospital for Women, a charity hospital
in the Brookline neighborhood of Boston for poor women.* A stately
four-floor building that rises from the banks of the Muddy River, it
goes by simply "the Free." Surgeries take place in the basement. If Miriam is lucky, sometime in the next four hours Rock will open the doors
of the surgical ward and hand her a bloody tissue containing an ovary
he has just removed from a patient.

Miriam would immediately put the tissue in her jar, which was full
of a calcium solution she had devised to wash it in. Then, she'd run

---

* The founder of the Free drew inspiration and support from James Marion Sims.
Like Sims's Women's Hospital in New York, the Free served as both teaching hospital
and a convenient source of human subjects.

as fast as she could up the four flights of stairs to her lab on the third floor. At 18x magnification, she could see the honeycombed matrix of bulging, fluid-filled sacs that compose the ovary, each one a watery home for an immature egg. She would puncture each transparent sac, or follicle, and suck out the bright-yellow fluid with a fine glass needle. Then she'd let the contents slip back out onto the glass dish and hunt for the precious tiny specks. One of them, she hoped, would contain her prize: a ripe egg.

Although it's the largest cell in the body, the human egg is a tiny thing, no bigger than the period at the end of this sentence. Most people need a magnifying glass to see it. Not Miriam. She could pinpoint an egg with her naked eye, could even tell by looking whether it was deformed or normal. She'd taught herself to do so through years of hunting. She would spend hours in the lab, picking apart follicles under the microscope. It was tedious work, "a delicate and frustrating business"—and she loved it. "She was a scientist, with a scientist's mind, and a scientist's precision, and a scientist's belief in the importance of following protocols," says Margaret Marsh, a historian at Rutgers University and coauthor with her sister Dr. Wanda Ronner of the 2008 Rock biography, *The Fertility Doctor: John Rock and the Reproductive Revolution.*

According to Miriam's charts, the woman who provided this ovary was thirty-eight years old and married, with four children. She was admitted to the hospital with a diagnosis of "relaxed perineum, lacerated, cystic, eroded cervix, and prolapse," meaning her uterus had collapsed and was sagging into her pelvis. Like most women facing a problem with their reproductive organs, she had found her way to Rock, who was considered the most capable fertility doctor in the country. In this case, he removed not only the woman's uterus but also her right ovary and Fallopian tube, free of charge. She provided a service for him too: she was one of 947 women who agreed to let him use her reproductive cells and tissues in his research. He made sure to perform the surgery on the tenth day of her menstrual cycle, when she was due to release an egg.*

Miriam pierces a single follicle, bluish in color and the size of a small hazelnut, and transfers the fluid to a glass dish with a flat bottom known as a watch-glass.† Within it is an egg surrounded by a halo of tiny orbs known as granulosa cells, which secrete hormones and cushion the delicate egg from the outside world. She washes it in a solution of salts and transfers it from the watch-glass to a thin-necked glass known as a Carrel flask. For the next twenty-four hours, she'll leave it in an incubator with some of the woman's blood serum at 37.5 degrees Celsius to mature. At this point, the egg releases a tiny proto-cell known as a polar body, which signals that it is capable of being fertilized by a sperm.

Sperm, of course, is the other half of this equation. It had to be fresh, no more than a few hours old. It was typically provided by male medical students, who were paid five dollars for each "specimen." To fetch it, Miriam would dash over to the hospital after midnight, long after her children were in bed, running the three or four blocks from her apartment to the Free. Wednesdays were when she introduced a newly washed egg to a cloud of sperm solution in a watch-glass, and prayed that two would become one.

This was the part she loved, the part she never got tired of. She'd watch the dance through her microscope, the sperm thrashing their tails, cartwheeling and cavorting their way toward the egg. Their force seemed to spin the egg round and round in a microscopic waltz. It was "the most fascinating thing to see those sperm—they're so active, they twirl the egg around, like nothing I've ever seen that's really as thrilling

---

* It's unclear how much the woman was told about how her reproductive tissues would be used. For Miriam's other project, the embryo study, she asked participants to keep track of their ovulation cycles and have unprotected intercourse days before surgery, explaining that this was important to help scientists "find out about the early stages of the baby" and ultimately help those with infertility conceive.

† In biology today, this term usually refers to lens-shaped glass with a convex bottom.

as that . . . just to sit and watch that and to think that this is going to be a baby," she would recall. Then she would go home and pray.

Rock had hired Miriam six years ago as a technician to work on what he called "the ova study." This was the first step in his plan to cure the scourge of infertility, which was at the time a complete scientific mystery. A practicing Catholic, he had stumbled into infertility research after a residency treating poor women in the slums of Boston. Their plight had opened his eyes to many forms of suffering: the inability to have children, and the inability to choose how many children to have. He particularly wanted to help women who had healthy ovaries but blocked Fallopian tubes, which were responsible for one-fifth of infertility cases he saw in his clinic.

If he could find a way to create embryos outside the womb, he could sidestep the tube problem and help these women fulfill their desire to become pregnant. "What a boon for the barren women with closed tubes!" he wrote in an unsigned editorial in the *New England Journal of Medicine* in 1937.

Miriam, too, felt for those "barren women," and was proud to contribute to a technology that might someday help them become mothers. But her real passion was solving the scientific riddle of fertilization outside the womb. For her, the ova study represented the fulfillment of a career derailed, the chance to be a part of a larger scientific project. She was in awe of Rock, a stately gentleman with silver hair and an air of priestliness about him. For the thirty years she knew him, he called her "Mrs. Menkin," which she hated, but she never once corrected him. She referred to herself proudly as his "egg chaser."

Rock, a clinician, was the scientific mind who had designed the experiment. But the technical details were beyond him. Fortunately, the details were where Miriam thrived. She was "a member of that legion of well-educated women who, with job titles such as technician or secretary, were indispensable to medical research in the first two-thirds of the twentieth century," as Marsh and Ronner put it in

The Fertility Doctor. "Working behind the scenes, often barely acknowledged, many of them with PhDs themselves, these women took the only jobs they could get, typing the papers of male researchers or performing tedious laboratory experiments." Though her official title was assistant, Miriam was certainly one of those women who would today be recognized as a scientist in her own right, says Marsh: "She wasn't just somebody's gofer."

Rock was always in a hurry, too focused on delivering babies or performing surgeries to keep a close eye on the ova study. "Dr. Rock was never there—he didn't know anything about it," Miriam recalled. She was on her own. She designed all the protocols herself, teaching herself to wash the egg three times, and expose the sperm and egg to each other for fifteen minutes to half an hour. Though she described herself as a "sloppy technician"—and, once, "the rottenest technician you can imagine"—her notes reveal her to be meticulous to the point of perfectionism. Each time she ran the experiment she would change some small factor, like the type of culture or the courtship intervals for sperm and egg, to see if it would change her results. It never did.

After six years with no results, the shine was coming off the exciting task that lay before her. She had examined around 800 eggs, and attempted to fuse 138 with sperm. Some days, she felt like giving up. She hated to take her $1.25 an hour, she said, because she didn't want scientific funds from the Carnegie Institute of Washington, which paid part of her salary, to be wasted. But she trudged on. Every Tuesday, she hovered outside the OR waiting for her prize. On Wednesdays she mixed an egg with sperm, and prayed. On Fridays she came into the lab with bated breath. And for the past six years, every Friday, she dug out the egg from its flask and peered into that glass and saw—nothing. Just an egg, and a bunch of dead sperm.

This week, something was different.

The "something different" was named Lucy Ellen Menkin, and she was eight months old. That week in February 1944, Lucy—"an *in vivo* specimen," as Miriam liked to say—had just cut her first tooth. For two nights straight, Miriam had stayed up all night with the nurse, soothing her as she cried. Miriam's husband, Valy, a research pathologist at nearby Harvard University, wasn't much help. He generally left childcare duties to Miriam, and anyhow, he was still peeved that she had gone off to work for Rock in the first place—before that, she'd been his unpaid research assistant. He particularly resented that she worked odd hours: she often went into the lab at two a.m. on Wednesday nights to fertilize eggs or fetch sperm.

Miriam went anyway. "It was just something I had to do, and it was the only interest I ever had," she said.

In 1944, Miriam was forty-two and a mother of two. The few surviving photos of her show her as maternal, beatific. In one portrait she gazes down at her young son, Gabriel, while her arm rests on her husband's shoulder, a serene smile playing on her lips. Her straight dark hair is parted down the middle, and she wears a checkered jacket and dark shirt with a deep collar. In her letters, she refers to wearing the same clothes over and over; later in life she had one black dress, and never had the money to buy a new one. She worked by the hour; no benefits, no vacation. When Lucy was a newborn, she would run home in the middle of the day to breastfeed her, and then come straight back.

For what she described as "innumerable" hours of work she got a check for $92 a month, which she handed over promptly to her baby's nurse for her care. Meanwhile, Valy controlled all her other finances, giving her a meager weekly allowance—$35—for grocery bills and housekeeping.

As a child, Miriam dreamed of becoming a doctor like her father, Dr. Friedman, who had practiced medicine in Chedrin, Latvia. "A real country doctor," as she called him, his patients came in horse and buggy, and paid him in buttons. When the Friedmans moved to the

United States in 1903, he became a specialist in internal medicine. He made enough money in New York City to provide Miriam a comfortable upbringing with household help, leading her to quip that she never "learned to cook or sew or do anything useful." When she was eight, she would listen spellbound as he told her how science would soon come up with a cure for diabetes. Before she was born, her mother had given birth to a boy who had died as an infant. "I was a substitute son and his greatest desire in life was for me to go into medicine and take over his practice," she recalled.

Miriam took steps toward that goal, graduating from Cornell University with a degree in histology and anatomy in 1922. The next year she earned her master's degree in genetics, and taught biology and physics in New York. But when she decided to follow in her father's footsteps, she ran up against her first obstacle. She was rejected by two of the top medical schools in the country, Cornell and Columbia. The reason was almost certainly due to her gender. In the 1920s, most medical schools did not admit women as a rule; only the occasional female student was accepted on a case-by-case basis. It was only in 1945, after facing a drastic shortage of applicants during World War II, that Harvard Medical School would begin admitting women as official policy.

Whatever the reason, the rejection would sting for the rest of her life, an undercurrent of inferiority that raised its head at inopportune moments. Miriam became obsessed with the fact that she could have been a researcher with a degree, working on her own project. Every June, she would read the names of medical school graduates in the newspaper, torturing herself with the fact that she could have been one of them. It would also shape her research path. Without an advanced degree, she could not pursue her own scientific interests, and instead worked mainly as an assistant helping others pursue theirs. "I really was nobody. If you don't get a doctorate in this kind of field, you work under people," she once said. "You're just in a different category."

With few prospects of becoming a doctor herself, she married a medical student instead. Most teaching jobs and research labs excluded married women, so Miriam worked as a secretary to help put her husband through his courses, going so far as to get another bachelor's degree in secretarial studies. She never lost sight of her goal to get an advanced degree. In the late 1920s, she managed to take the first two years of pre-clinical sciences at Harvard, as well as a course in bacteriology and embryology. But because she didn't have the $200 to pay tuition, she never received credit. Rock would later marvel that she was the only person in history of Harvard Medical School who took bacteriology and embryology without credit.

She also served as an assistant in her husband's lab at Harvard. It was there that she ran into Dr. Gregory "Goody" Pincus, a young Harvard biologist who had overlapped with her as a student at Cornell. Now he was one of those researchers she envied, a scientist in his own right. Pincus would go on to become, with Rock, the co-creator of the Pill. But at the time, he was engaged in a rather different endeavor. He had just gained notoriety as the Frankensteinian scientist who had fertilized rabbit eggs in vitro, implanted them back into the mother, and raised them to healthy, hopping adulthood. Now he needed an assistant.

Miriam gladly took the job, putting into motion a lifelong career studying germ cells. Her task was to extract two key hormones from the pituitary gland of a sheep, follicle-stimulating hormone (FSH) and luteinizing hormone (LH). Mixing them together, she injected the slurry into the uteruses of female rabbits to make them ovulate extra eggs, known as "superovulation."* Miriam was thrilled with the work. Wouldn't it be wonderful, she once told Pincus, if these extracts could be used on infertile women to help them ovulate? It was around that time that she became pregnant with her own son, Gabriel, an occur-

---

* Today, women going through IVF are often treated with similar hormones to stimulate their ovaries to grow extra eggs, a process known as controlled ovarian hyperstimulation.

rence she always attributed to an accidental ingestion of that "potent extract" in Pincus's lab.

Her pregnancy, twelve years into her marriage with Valy, was bittersweet. Secretly, Miriam had been saving up money to divorce her husband. When she realized she was pregnant, she felt that option was off the table.

Pincus was ahead of his time. In the 1930s, creating life in a dish raised the specter of scientists playing God. Reporters would compare his achievement to the techno-dystopia of Aldous Huxley's *Brave New World*, where genetically modified babies were made en masse via test tubes. A *Collier's* magazine article predicted a world in which men were no longer necessary, "where woman would be self-sufficient; man's value precisely zero." The *New York Times* speculated that in vitro fertilization would lead to the rise of "human children being brought into the world by a 'host-mother' not related by blood to the child"—and that such a technology would allow certain women "of special aptitudes" to give birth to dozens of children by hiring surrogates.

His work was too controversial, his comments too brash for Harvard. "Pincus did not seek publicity," as his biographer, Leon Speroff, put it; it "found him." In 1937 he was denied tenure. He would soon return to England, and Miriam would lose her research position. For a while she worked odd jobs, including a short stint at a state lab in Boston, where the research goal was to develop a new test for syphilis. Her job was to drag rabbits from their cages and inject them with the disease. It was too much; she quit after three weeks.

When she heard of an opening in Rock's lab, Miriam was immediately intrigued. "If you think my services may possibly be of use to you I should of course be happy to have the opportunity of seeing you at your convenience," she wrote. She included her résumé, which showed that she had taken all the pre-med requirements. This was exactly what Rock was looking for: someone who had been educated in the relevant scientific disciplines, who had the technical skills to unite egg and sperm, and who had the patience for the detailed and painstaking work he lacked.

At her interview, Miriam was intent on impressing Rock. She told him of her cytology coursework with the influential geneticist E. B. Wilson at Columbia. He wasn't moved. Her work in Pincus's lab, however, did the trick. "I told him that I had once worked for Dr. Gregory Pincus, though in a very minor capacity," she recalled. "I used to prepare those pituitary extracts: FSH and LH, for Dr. Pincus's experiments on rabbit eggs. SUPEROVULATION! that one word did it . . . I was hired." The following Monday, she joined the hunt for human eggs.

By the time she succeeded, in 1944, the political tide had changed. America was deep in the throes of World War II. Science and technology were seen as central to the war effort and the race to stay ahead of the Germans. Technologies to enhance fertility, including artificial insemination, were seen as a way to restore the generations of dead Europeans after the war. No longer was the possibility of test-tube babies likened to Aldous Huxley's techno-dystopia, or reason to fear men becoming obsolete. Now the promise of a solution for infertile women seemed more hopeful than horrifying. This "scientific affront to womanhood," as *Time* put it, proved that with science, man had at last conquered nature.

———

The discovery that would upend reproductive science happened on a Wednesday. Miriam left Lucy with the nurse as usual. She walked the three or four blocks from the "lonely part of Brookline" where she lived to the Free in a sleepless daze. Because she was so tired, she made a mistake. She washed the sperm only once, instead of her usual three times. Then, when she introduced the sperm and egg, she let her guard down again. Losing herself in the familiar dance, she felt her eyelids grow heavy, and she dozed off over the microscope. By the time she looked back at the clock, an hour had passed—twice as long as her normal protocol.

Years later, in her typical self-deprecating style, she recalled what had transpired: "I was so exhausted and drowsy that, while watching under the microscope how the sperm were frolicking around the egg, I forgot to look at the clock until I suddenly realized that a whole hour had elapsed . . . In other words, I must admit that my success, after nearly six years of failure, was due—not to a stroke of genius—but simply to cat-napping on the job!"

On Fridays the lab was mostly empty. Miriam came in as usual around ten a.m. When she opened the incubator, the steam had evaporated to the top of the flask, obscuring her vision. She worked with a lighted match, cutting through the vapor with the flame. She transferred the egg into a watch glass, looked through the microscope—and gasped. The cells had fused and were now dividing, giving her the world's first glimpse of a human embryo fertilized in glass. Miriam blinked, and looked again. "When I realized what it was, I nearly dropped dead," she recalled. She screamed; where was Dr. Rock?!

Miriam took the elevator down, trembling, and bumped into her colleague, pathologist Dr. Arthur Hertig, who rushed upstairs with her to confirm her finding. As the lab filled up with onlookers—"everybody came running in to look at the youngest human baby that had ever been seen"—Miriam kept the egg in her view. "And there I was sitting there in their midst with this beautiful two-celled egg," she recalled. Finally, someone was able to get a hold of Rock, who was delivering a baby in a Boston home. "We telephoned him . . . When he saw what was in the dish, he became pale as a ghost," she said.

To preserve the egg, she had to remove fluid from the dish and replace it with a fixative, drop by drop. For hours she worked the egg, eating a sandwich with one hand, dropping fluid with the other, long into the night. She was "afraid to let out of my sight that precious object, to achieve which had been an unfulfilled dream for 6 years," she would write in a lecture years later.

Miriam and the other researchers began to argue whether to put the egg back in the incubator, or preserve it—and if so, which method to use. In the midst of the debate, she forgot to have the egg photographed. When the argument was over and a decision reached, she went back to the microscope to begin the preservation process. The egg was gone. It would be "the first miscarriage in a test tube," she remembered ruefully.

She would spend hours looking for that two-celled egg, drawing fluid with her glass pipette, to no avail. But now that she knew the method, she could do it again. The following week, she fished out twelve eggs from the ovary of a thirty-one-year-old woman who had also had her uterus removed. Using the same method, she managed to create another two-cell egg, and two three-cell eggs from her and one more woman. This time, she remembered to photograph them. In the published photos, it's easy to make out the dark pinheads of sperm surrounding the egg; one has clearly made it into the zona pellucida, the jelly-like coating that encases the cell. Rock and Hertig immediately sent the images to the Carnegie Institute.

They were "our pride and joy," Miriam said. The photos proved "without a doubt—we had the egg."

Rock and Miriam reported their findings in a brief inital report in *Science*, with Miriam as Rock's coauthor. Within hours of its publication, every wire service and major newspaper in the country had picked up the story. Here, at last, was proof that a cure for infertility was on the horizon. Rock had designs on bringing an embryo to the four-cell stage, then eight cells, and then—who knows? "Whether we shall be able to fertilize human egg cells for the practical purpose of reimplanting them in a mother's body remains to be seen," Rock told a reporter. "It certainly is not beyond the realm of imagination, and it seems to offer about the only hope for women whose tubes have been destroyed."

Their achievement was also an early glimpse into the formation of human life. "Dr. Rock's experiment is the first ray of hope ever held out to the thousands of women who are childless because of defective Fallopian tubes," wrote the magazine *Science Illustrated* in September 1944. "But the most significant aspect of Dr. Rock's success was the opening of vast new fields for research on the beginnings of human life. No one can say where it will lead."

Now, despite her lack of a degree, Miriam was poised to become a reproductive scientist who would push the frontiers of fertility even further. But then something happened that neither she nor Rock could have anticipated: her husband lost his job. Desperate to keep Miriam and repeat the experiments, Rock begged the dean of Harvard Medical School to let Valy keep his position, but to no avail. Decades later, Miriam would be interviewed about this part of her life by a reporter working on a biography of Rock. "That was really very sad—it was a great disappointment—after working so hard and giving up so much . . . " She trailed off. As his wife and mother of his children, she was left to follow him to Duke University, in North Carolina, where IVF was considered a scandal—and, to at least one doctor, "rape in vitro."*

Without Miriam's skills, no one in the Rock lab would definitively succeed in fertilizing an egg *in vitro* ever again. IVF wouldn't truly come to fruition until the birth of Louise Brown, the world's first test-tube baby, in 1978. Miriam, for her part, would spend the rest of her life trying to return to her research. "My life's ambition was to have a chance to repeat it," she said. She once said that she considered herself a failure as a wife and mother. She devoted herself to science, she added, because she wanted to get something right in her life.

Biologists today remain entranced by sperm. After all, says Dr. Scott Pitnick, a biologist who studies sperm evolution at New York's Syracuse University, "They are the only cells that are cast forth from the soma to

---

* The diaphragm, meanwhile, was referred to as "the instrument of the Devil."

spend their lives essentially as free living organisms in a foreign environment, which is the female reproductive tract." They also evolve lightning-fast, resulting in a breathtaking variety of shapes ranging from a scythe to a three-headed Cerberus. In many cases, "you can look at a single sperm cell and know the kingdom, phyla, class, order, family, genus, and species that it came from," Pitnick says.

Unbeknownst to them, they are all following in a rich tradition of male scientists waxing poetic about the products of their testes. "Sperm is a drop of brain," wrote the ancient biographer Diogenes Laertius.

Antonie van Leeuwenhoek, inventor of the modern microscope, also famously overestimated sperm's abilities. After peering into a drop of his own semen in 1677, the Dutchman fancied that he saw the whole of a human being within the head of a single sperm. "I have sometimes imagined, as I examined the animalcules in the male seed of an animal, that I might be able to say, there lies the head, and there, again, lie the shoulders, and there the hips," he wrote to the Royal Society.* Each sperm, he concluded, contained a tiny person, curled up and pre-formed, which merely took root in the female and unfolded. Meanwhile, Leeuwenhoek flatly rejected the notion that women produced eggs, calling the idea "addle-pated" and "entirely erroneous."

By contrast, the female germ cell is usually cast as the large, round, sedentary maiden who waits patiently for a prince to deem her worthy. She is, as a 1983 essay in *The Sciences* put it, "a dormant bride awaiting her mate's magic kiss, which instills the spirit that brings her to life."† In many a scientific textbook, the story of egg and sperm is little more than an exercise in Darwin-era gender stereotypes. This "imagery keeps alive some of the hoariest old stereotypes about weak damsels

---

* He hastened to add that he had produced his sample not by "sinfully defiling myself," but by "conjugal coitus," and that the Society should feel free to burn or publish his letter as they saw fit.

† To be clear, the essay was critiquing this idea, not supporting it.

in distress and their strong male rescuers," wrote anthropologist Emily Martin in her classic essay "The Egg and the Sperm." Worse, it naturalizes these stereotypes in the language of science.

In reality, the egg undergoes its own drama, complete with brutal competition and a daring quest. An egg is a moment stopped in time, halted just in the middle of cell division. It is Han Solo cryogenically frozen, waiting to leap into action. Each month after puberty, the body "chooses" a cohort of around twenty follicles to escape their state of suspended animation and ripen within their follicle sac. One egg eventually muscles its way out in front, its follicle swelling to up to 30 millimeters—a little over an inch—in diameter. Bulging from the side of the ovary, it releases hormones that feed back and inhibit its peers, causing them to shrivel up and die. The lone survivor is the Graafian follicle,* the one destined to explode through its follicular sac in a burst of hormonal glory known as ovulation.

The egg now hangs in the balance between the ovary and the delicate mass of webbed fingers, called fimbriae, that coalesce at the tip of the Fallopian tube. Like a Venus fly trap, the fimbriae stiffen, fill with blood, and delicately pluck the egg into the tube. Contractions of the tube and minuscule hairs called cilia draw the egg farther down, like a tiny crowd surfer. But the egg isn't just pulled along by outside forces. Some force within it guides it as well, captaining it to its destination.

Biologists today consider sperm not as independent explorers of uncharted territory but as one half of a conversation. These cells are mutually synergistic partners; neither can do its job without the other. Take, for example, the overlooked importance of female fluids, which are essentially an extension of the female reproductive tract that accompanies the eggs. That includes vaginal fluids, ovarian fluids, and Fallopian-tube fluids that help trigger "capacitation," the set of chemi-

---

* Named for the Dutch anatomist Regnier de Graaf, who actually wasn't all that bad, as we'll see in Chapter 6.

cal changes that prepare sperm to swim up the tube and enter the egg. When they capacitate, sperm lose the protein helmet around their head, revealing receptors that sniff out chemicals in the female reproductive tract and lead them swiftly to the egg.*

In 2020, Pitnick published a paper on the significance of these female-triggered changes in sperm. Capacitation is so crucial, he wrote, that "the goal of achieving fertilization *in vitro* in a mammal was stymied" for decades as researchers struggled to figure out the missing ingredient. The breakthrough, according to him, came in a pair of studies on rabbits and rats in 1951. When I pointed out to him that he seemed to be overlooking Miriam's achievement of human IVF as early as 1944, he was taken aback. "I'm a bit embarrassed," he said, admitting that he'd never read the Menkin paper—which speaks more to her erasure from IVF history than his knowledge of the scientific canon. (After our conversation, I sent it to him, and he thanked me.)

His best guess, he said, is that the blood serum she washed the egg in somehow triggered capacitation without her realizing it. "I can't imagine how else it could have worked," he said.

When the sperm arrives at its long-sought-after destination, it encounters the corona radiata, a ring of orbiting cells that surround the ovum and are named for the fact that they resemble a regal crown. These swell the size of the egg, which would otherwise be utterly microscopic, to an orb just big enough to see. Corona cells are the bouncers of the egg cell, judging and filtering out sperm, deciding which get in and which are so out of luck. Textbooks typically report that the fastest, strongest sperm "wins." But that explanation is far too simple, says Dr. Kurt Barnhart, a professor of obstetrics and gynecology at the University of Pennsylvania Medical Center. "I don't think we understand why that one was selected and the other one next to it was not," Barnhart says. Instead, it's likely that both cells play an equal role in which sperm gets in.

---

* In other words, without female help, sperm are literally incompetent.

Next is the zona pellucida, the jelly-like membrane that encases the egg itself. Although sperm, to Miriam's eyes, appeared powerful enough to twirl the egg around and penetrate it, in reality, they flail far too weakly to do any such thing. "The mechanical force of its tail is so weak that a sperm cannot break even one chemical bond," wrote Martin, the anthropologist. Fortunately for them, this membrane is sticky, covered in tiny sugar chains that capture sperm and plaster them to the egg's surface.

Once a sperm touches the zona, the egg begins a remarkable cascade of changes. First, it rapidly shuts down the remaining sperm receptors on its surface to block stragglers from entering. Then it releases granules of calcium that harden the zona, turning it from flesh to skeleton, invitation to barrier. At this point the corona dissipates, falling off like a crown dissolving. Within twenty-four hours, cell division begins. On the fifth day following conception, the embryo hatches from its shell and implants into the tissues of the uterus. At this point, the mysteries only multiply. "We really don't understand how it moves down the tube, and how the cells change to placental cells, or this cell becomes the brain, or this cell becomes the body," says Barnhart. "It's really quite amazing."

Here's another thing that's amazing: All those traits that biologists like to praise about sperm—that they're sleek, decisive, powerful—are likely products of the biology of the egg. When we talk about sperm evolution and behavior, what we often fail to recognize is that the egg and its surroundings are involved in every step of the process. "If I had to wager, it's that female reproductive-tract traits are evolving independent of ejaculate traits," Pitnick says. "Males are just struggling to keep up." Just as with vagina science, when we overlook the contribution of the female, we miss out on understanding the delicate biochemical dance that is fertilization, and the interplay of male and female forces that go into it.

The first person to lose themself in the reverie of egg and sperm was Oscar Hertwig. A German zoologist who lived in the second half of the nineteenth century, the object of Hertwig's fascination was not a human egg and sperm but the germ cells of a purple-spined burr that waddled along the edge of the Mediterranean Sea.

For centuries scientists had longed to see the climax of life-making: the fusion of egg and sperm. Their repeated complaint was the opacity and impenetrability of the female body. Scientists had long tried to get inside her, but in vain; the fertilized egg was nowhere to be found. With no evidence to go on, a plethora of unfortunate ideas arose to explain conception, including that the embryo was a "cooked" mixture of semen and menstrual blood (that started with Aristotle, who noted that menstrual blood must serve some function, and pregnant women do not menstruate), or that both sexes produced semen during intercourse. In Hertwig's time, one popular view was that the sperm transmitted a subtle mechanical vibration that stimulated the egg to begin developing.

But Hertwig didn't think so. If someone could see the moment of fertilization for himself, they could put these silly theories to rest.

It had taken centuries to get to this point. In the 1600s, William Harvey, the British physician who discovered that the heart is a pump, became the first to champion the existence of the egg. "All animals whatsoever . . . nay, man himself, are all engendered from an egg," he declared in his book *Disputations Touching the Generation of Animals.* On the title page, the Greek god Zeus holds a giant egg above his throne, spilling forth all manner of living creatures. The egg bears the inscription *ex ovo omnia:* all life comes from an egg. The problem was, Harvey didn't have proof.

Two hundred years later, in 1827, Karl Ernst von Baer, a biologist working at the University of Konigsberg in Germany, became the first to finally catch a glimpse of the egg herself. Baer was convinced that mammalian ovaries produced eggs. He believed he had seen them in the uteruses of dissected dogs, "formed of such very delicate membranes that a breath could change their shape, almost as those soap bubbles which children enjoy blowing through the air." He decided to cut open a dog that had recently mated, in the hopes that her follicles would have just opened to release their eggs. Miraculously, his colleague had just "such a dog in his own home," he recalled. "She was sacrificed." After dissecting the "still vital uterus," he turned his attention to the ovary.

And there it was, "so plainly that a blind man could scarcely deny it."

"When I observed the ovary," he wrote, "I discovered a small yellow spot in a little sac. Then I saw these same spots in several others, and indeed in most of them—always in just one little spot. How strange, I thought, what could it be? I opened one of these little sacs, lifting it carefully with a knife onto a watch glass filled with water, and put it under the microscope. I shrank back as if struck by lightning, for I clearly saw a minuscule and well-developed yellow sphere of yolk." What he saw looked strikingly like a chicken egg. Rather than perfectly spherical, the yolk was a bit squashed, with a halo that reminded him of the planet Saturn. He called it *ovulum*, or little egg.

The egg, though, was just one half of the equation. That's where Hertwig came in. Born in Hessen, Germany, Hertwig was a student of Ernst Haeckel, the naturalist (and eugenicist) who taught at the University of Jena in Germany. Haeckel convinced Oscar and his brother Richard to give up chemistry and take up medicine, and the pair soon set out to understand how animals develop from embryos. Bald and fastidious, with a neat triangle of a beard, Hertwig wanted to understand the mysteries of early life. In 1875, he learned that his brother was taking a research trip with Haeckel to a laboratory on the Mediterranean Sea. Hertwig was twenty-six and had just received his medical degree; he had also accepted an assistantship at the University of Bonn. He immediately resigned his new position to join them.

And it was there, in a seaside laboratory in the Bay of Naples, Italy, that he discovered his true love: the sea urchin.

Sea urchins, or *Toxopneustes lividus*, were Hertwig's ideal study animal. Fast-growing and plentiful, they spawn readily and on command. One squirt of potassium chloride and the males lose their loads, floating dense clouds of millions of sperm into the waters. Masses of urchins huddle together for love-making, and egg and sperm cloud the waters around them. A single female sea urchin can spawn millions of eggs at once; when fertilized, they all begin dividing in sync. Best of all, their gametes are utterly transparent, allowing any observer to peer into the depths of their glass jelly and see the magic going on within. For Hertwig, these purple-spined burrs of the sea were a scientific godsend.

Armed with only his microscope and a smear of sea-urchin jizz, Hertwig set out to catch an urchin in flagrante. One spring day in 1875, he did. Peering at a clear urchin egg through his lens, he homed in on its nucleus, a dark spot in a pool of jelly. Then, he dropped some sea-urchin semen near the egg. Before his eyes, a tiny, wriggling sperm cell slowly approached the surface and climbed up against it. Then the sperm's nucleus appeared—*within the egg*. He watched the nucleus of the sperm make its way toward the nucleus of the egg, as if drawn by a magnet. Suddenly the two fused into one. A few minutes later a membrane had formed around the egg, and one nucleus had become two.

Hertwig had just become the first to witness the earliest moments of fertilization. Through a globe of jelly as clear as glass, he had watched life unfold. A "lofty and lonely" man in the words of his biographer, Hertwig nonetheless felt the poignancy of witnessing such a momentous union. The emergence of a single nucleus where there once were two, Hertwig wrote, "arises to completion like a sun within the egg."

After that, it was no longer tenable to argue that either the male or female contributed everything to conception. Clearly, both nuclei fused into one and then divided, suggesting that both were necessary to the endeavor. The advent of better microscopes would allow scientists to probe the contents of egg and sperm to determine what exactly the stuff of this dual inheritance was—the coiled threads that would come to be known as chromosomes. Within twenty years, science determined that each germ cell contributed an equal number of chromosomes to the resulting offspring. One protégé described Hertwig's observation as having, "with a single brilliant stroke, illuminated the field."

"The nuclei of the two germ-cells are in a morphological sense precisely equivalent," wrote E. B. Wilson, the geneticist who would one day become Miriam's teacher, in 1895. "The two sexes play, on the whole, equal parts in hereditary substance."

—

In 1956, Miriam was invited to give a talk on her egg research to a classroom full of schoolchildren. In it, she marveled at the process by which egg meets sperm. "When you think how small this egg is, does it not seem amazing that once it is liberated from the follicle, and falls into the comparatively tremendous body cavity, that it does not get lost?" she asked the children. "How does such a tiny thing, a little speck, find its way to the place where it is supposed to be going?" In a similarly wondrous way, Miriam would eventually find her way back to Rock and the world of fertility research. In the case of the egg, marvelous changes in the female body propel it forward. In the case of Miriam, it was sheer persistence and a bit of luck.

Everywhere her husband's job took her, she sought out opportunities to chase down eggs and lab space. Once she and the family had settled in Durham, she started knocking on doors of reproductive researchers who might provide her with ovarian tissue, and asked Rock to recommend her skills. "It took someone who was very sort of not shy and persistent to be able to approach people out of the blue and say, 'Hey, I used to work with John Rock, could I have some lab time?'" says Sarah Rodriguez, a historian at Northwestern University who has written about Miriam's contributions to reproductive science. "That takes a certain amount of—if not boldness—at least confidence and desire."

As a wife and mother, her opportunities were limited. A surgeon at Duke Hospital, Dr. Hamblen, offered her a job in a fertility clinic. The job turned out to be mainly secretarial—"a dirty trick," she recalled—working on a bibliography for his book. Hamblen told her she could chase down eggs in her spare time, which meant after five o'clock and on weekends. But, he added, the surgeons at Duke would likely not cooperate and it would be a waste of time to bother with. That was "ridiculous," she recalled, "because you couldn't do that sort of work after five." She had two small children and still hadn't found a permanent place to live, due to the housing shortage after World War II. She wrote to Rock that "the prospects here as far as doing any ova work are still somewhat discouraging."

After two years, Duke decided not to extend Valy's appointment. The Menkins moved again, this time to Philadelphia, to a run-down house near Temple University's medical school, where Miriam's husband started a new job. Miriam had to start over. The only place she definitively found laboratory space was at Lankenau Institute in Philadelphia. "I think I shall enjoy working here," she wrote to Rock on April 12, 1947. The only "trouble" was their lack of research funds—they could not offer her a paid position, only lab space. She would have to find part-time work, to cover Lucy's kindergarten expenses.

A year later she was still there, working without funds. "I still haven't found any way to finance myself, and it would be nice to have reprints to include in any application for funds," she wrote to Rock in 1948. Five years after leaving the Free, nothing had materialized.

Throughout it all, Miriam continued her work for Rock remotely. Due in part to her home situation and in part to her perfectionism, it took her four more years to write the complete version of her first brief

IVF report. Every day when Lucy took her nap, Miriam would take two buses to the library, which was an hour away, and work on the paper. In 1948, she and Rock finally published the full version of their first IVF achievement in the *American Journal of Obstetrics and Gynecology*, with Miriam, at Rock's urging, listed as first author. It would become her greatest legacy, something she would recall with pride well into her eighties. "I always felt that I should pay Dr. Rock instead of him paying me because of all the fun," she once said. "It was the greatest thing that ever happened to me, being connected with him."

Over the next five years, the thought of returning to work with Rock was what kept her going. But to get back to him, there was one last thing she had to do.

—

The final stroke came on September 30, 1948. That night, Valy sat across the dinner table from her, berating her in front of her children. She had heard it all before. Claims she was planning to kill him. Threats to take both children away from the house. Valy was known for his temper and for getting into arguments at work. If he couldn't find someone to argue with at the lab, he'd come home and take his anger out on her. When he got into one of these fits he became an avalanche, an unstoppable force that had to run itself out. The best way to weather it, Miriam knew, was to say nothing, just endure.

But that night, she made a mistake. To distract him, Miriam suggested he call up his parents in New York—and let slip that she had called them herself. She'd been communicating with them in secret, begging them to convince him to come up to New York on weekends so she could have peace. She'd even been forced to go to them for money when he withheld her weekly allowance. Now Valy was implacable. "Children, your mother is a dirty stinking rat," he yelled. "She is a murderer and a liar." Miriam, as usual, stayed silent. She said only to the children, "Your daddy doesn't feel so well tonight."

In an attempt to escalate the situation further, Valy had promised to take Lucy to the movies after dinner, meaning they would get home after eleven. Lucy, who was five years old, suffered from severe seizures and behavioral problems. This required Miriam to spend much of her time in and out of hospitals, meeting psychiatrists and communicating with her teachers at school. Now Lucy was throwing one of her tantrums, screaming and kicking. Her hopes had been raised only to be crushed, and now she was left howling and screaming. "You get out of this house, Daddy, we don't want you around here," she cried.

Valy was not finished. "You have to decide. Do you want to stay with Daddy or with Mama?" he said to her. Then he turned to his nine-year-old son, Gabriel, and asked him the same thing.

"Both," Gabriel answered, growing distraught. "Do I have to decide right away? I got to think this thing over."

Miriam felt a great heaviness take over her. Her sister Esther had recently suggested she get a divorce, and each parent take one of the children. Miriam had resisted the idea. "I believed it would create too much of a trauma . . . the stigma of divorce would be especially hard on Gabriel," she wrote to her parents-in-law. Now she realized that the children were already suffering. And so was she. "Words are inadequate," she wrote, to describe the experience of living with such a man. She would continue sending detailed descriptions of his abuse, and if that didn't work, she would take the next step. "I am going ahead with plans to gain help and legal advice as to how to protect myself from further insults and threats of physical violence," she wrote.

That year she obtained a divorce and took custody of Lucy. The pair moved to a small apartment with a kitchenette in Philadelphia, and scrimped to make ends meet. For Miriam, freedom was sweet. A prolific poet who was often scrawling ditties and submitting them for publication, she wrote a poem called THRESHOLD around this time:

The joy of new-found FREEDOM
From dreary household care!

I revel in its beauty . . .
A precious thing and rare.

There's dirty dishes waiting,
But I feel gay and free!
I'm stranded on my door-step . . .
I cannot find my key.

Yet divorce came with new problems. Though her marriage had compelled her to follow her husband wherever he got a job, it also gave her the income to allow her to pursue unpaid work. Now she had no steady work, and sole custody over a disabled daughter. While married to Valy, she had trouble finding lab space to do her research, as she was limited by the hours of childcare. Now it was out of the question. Valy hadn't changed; instead of housekeeping money, he withheld child-care payments from her, demanding weekly reports of Lucy's well-being.

It was Lucy, in the end, who brought Miriam back to Boston, and to Rock. Given her seizures and her behavior, schooling had always been a challenge. In 1952, Miriam learned of a suitable school called Cumberland, a "school for people with special children with special needs," as she put it. It was in Boston, and it was dreadfully expensive; she would receive bills for up to $375 per semester (around $3,800 today). But Miriam was desperate. By then she had custody of both children, who were eight and eleven, so they packed up and moved to Boston to enroll them both in school. It was a dark time. Miriam was in a "bad, depressed state," and frequently fell ill. Money was a constant problem.

Miriam's life had not gone according to plan. She was forty-eight, a single mother of two, and hadn't had a regular salary since she left Rock in 1944. But despite everything, she had never given up her dream of attaining a doctorate degree. Soon after they arrived, she signed up for a class on cell staining. The course met twice a week, and she planned to take it during the day while the children were in school. To apply for the class, she needed a professor to recommend her. The trouble

was, all of her old professors had died. So she called up the only reference she had left: Rock. Her boss was shocked to hear from her on the
phone—and living in Boston, no less!

Without meaning to, Miriam poured out the whole story: how
her family had split up, how she was struggling to make ends meet,
how much she longed to return to research. Rock didn't hesitate. He
offered her a job, if she would give him just a week's notice to get the
funds together.

"Mrs. Menkin, why don't you come back to me," he said. "I want
you to come back."

———

Much had changed in the intervening years. Now the reproductive task
of the hour was not to make more babies but to prevent more babies
from being born. Rock had his own reproductive clinic, and his new
mission was to help develop a convenient method of contraception, an
undertaking that would lead to the approval of the Pill for contraceptive use in 1960.*

As Rock moved closer to his ultimate goal, Miriam worked behind
the scenes as his "literary assistant." She coauthored papers investigating whether women's menstrual cycles could be stabilized by light, and
whether a heating jockstrap could render men temporarily infertile.
She also looked into the possibility of freezing sperm at subzero temperatures, something she viewed as "a project of increasing potential
significance in view of the ever-present threat of nuclear warfare."

Miriam never returned to IVF. But she kept newspaper clippings
of relevant advancements—including the birth of Louise Brown—and

---

* To gain FDA approval, Rock and Pincus tested the Pill en masse in Puerto Rico,
seeking out women who were poor, lacked education, and needed a way to limit their
family size. Despite the fact that the tablets were experimental and came with severe
side effects, participants were told only that they would prevent pregnancy.

attended graduate courses to keep up with the evolving science of her field. David Albertini, an oocyte expert who attended Harvard as a graduate student in the 1970s, remembers that she would come to his seminars on human reproduction every week and sit in the back of the class. "She was just this humble, humble person," he recalls. "In the wintertime, she would come in with several woolen scarves, thick scarves covering her head. I mean, she would look like a peasant in the fields . . . But the few people who were there were like, This was Rock's technician. This was the woman who worked with Hertig to collect these embryos."

In modern times, IVF has taken off beyond either Miriam's or Rock's wildest dreams. No longer is it a solution only for "barren women with closed tubes." Now, combined with egg freezing, surrogacy, and the ability to inject a single sperm directly into the egg, it can also solve parenthood woes for couples with myriad other fertility problems, women who undergo cancer treatment, or those who want to give birth later in life. As of 2018, more than 8 million babies had been brought into being using IVF.

The technology that Miriam laid the groundwork for has also expanded the notion of parenthood in a more fundamental way. In 1981, when Elizabeth Carr, the first US "test tube baby" was ushered into the world, it was during an era of conservative backlash. Ronald Reagan was president; the field of American IVF had been hamstrung by debates over the ethics of abortion and a federal moratorium on funding embryo research that followed in the wake of *Roe v. Wade*. In this environment, the main criteria for prospective IVF parents, one reporter wrote, was "youth (under 35), good health, bad tubes, and a husband"—plus the equivalent of $13,000 in today's dollars.

Today, IVF allows an ever-widening array of people the option to give birth biologically: single mothers, queer parents, asexual people, and more. In practice, of course, reproductive technologies are shaped by the same inequities as the rest of American healthcare. While IVF was a "profoundly revolutionary technology," says Rene Almeling, a Yale sociologist and author of *Sex Cells: The Medical Market for Eggs and*

Sperm, "it remains an expensive medical technology in a society deeply riven by inequality." White, married, heterosexual parents still make up the vast majority of those who use reproductive technologies, which are often not covered by health insurance.

The landscape of reproductive technologies is still haunted by questions of eugenics, commodifying bodies, and playing God.* But what can be said is that the birth of IVF, along with the contraception pill—the vision Rock would help bring into being in the 1950s, again with Miriam's help—changed the meaning of human reproduction forever.

Perhaps no one knows that better than Elizabeth Carr. Now a patient advocate and a forty-year-old mother herself, Elizabeth was well aware of what her birth meant to couples who badly wanted to have children—like her mother, Judith, a twenty-eight-year-old schoolteacher in Massachusetts, and her father Roger, an engineer. Judith lost both Fallopian tubes after three ectopic pregnancies and was told by a doctor she would never give birth. Yet her doctors, Drs. Howard and Georgeanna Jones, successfully fertilized one of her eggs with her husband's sperm and transplanted it back into her uterus. Elizabeth likes to say she was famous since she was three cells old, when she was photographed *in utero*; at age one she graced the cover of *Time*, round and rosy-cheeked, clutching a petri dish in front of a microscope.

Elizabeth grew up steeped in the history of reproductive technology, able to rattle off the elevator pitch of how IVF worked since the age of six. ("A sperm and an egg are fertilized in a petri dish, and once it's fertilized, it's put back in the mother, and nine months later you have a baby.") Growing up, she was on a first-name basis with all the major fertility experts at the time—Landrum Shettles, Edwards and Steptoe in the UK, and of course her own doctors, the Joneses. But one day, during one of the many reproductive technology conferences she went

---

* The Catholic Church, notably, still opposes IVF on the grounds that it replaces the marriage act and results in wasted embryos.

to as a teen, she saw a name she didn't recognize. It was at the bottom of a research poster on the early studies of fertilization: Miriam Menkin.

"It was literally just a footnote," she recalls. It stuck out to her not only because it was a name she didn't know, but because she almost never saw women in the field besides Dr. Jones. "Dr. Georgeanna, my doctor, was really a pioneer," she says. "And so to hear of another woman that was doing this cutting-edge stuff was fascinating."

Later, she looked Miriam up online and learned about her journey to unite egg and sperm. That story reminded her of what her parents had gone through to conceive: even by the late 1970s, nobody knew what protocol would work best to unite egg and sperm. All ten couples at the Joneses' experimental IVF clinic were undergoing slightly different protocols using different forms of hormonal stimulation, in the hopes that one would work. "My parents always say that they feel like they hit the lottery," she says. "They just kind of got lucky that their exact combination worked." It was exactly how Miriam would describe her achievement.

To see Miriam Menkin's life work, you have to fill out a form asking whether you are a registered sex offender, have ever been convicted of a felony, and are a US citizen. Assuming you answer correctly, you can enter the Forest Glen Annex, a gated 136-acre US Army installation in the leafy suburbs of Silver Spring, Maryland. This is the home of the National Museum of Health and Medicine, a brutalist beige building with opaline windows. The museum's varied and morbid exhibits include the skeleton of Able, a rhesus monkey who was one of the first primates to be sent into space, and artifacts from the assassination of President Abraham Lincoln, including fragments of his skull, his doctor's bloodied shirtsleeves, and the lead bullet that pierced his brain.

This military entity came to acquire Miriam's historic embryos through a circuitous route. In the 1930s, the Carnegie Institute was in an arms race. During the First World War, its Department of Embryology made it its mission to peer into the earliest stages of life, creating a vast collection of human and primate embryos to rival those of Europe. As microscope technology and techniques for fixing specimens advanced, Carnegie took the lead in the race to obtain younger and younger embryos. Its leaders were especially keen to obtain ones within the first two months of conception, to help science fill in details about this never-before-seen stage of development. The path of the fertilized egg, once shrouded in mystery, would soon be laid bare. *

The way to do that was through friendly gynecologists. John Rock, the preeminent fertility expert of his time, was key. In 1938, he and his colleague Dr. Hertig launched "the embryo study," a Carnegie-supported hunt for the earliest embryos. As his assistant, Miriam kept ovulation records of women scheduled for hysterectomies. With Rock and Hertig, Miriam helped Carnegie get its hands on thirty-four of the earliest embryos ever seen, representing the first seventeen days of life. Their images represented the first-ever visual record of early human development, and allowed artists to re-create almost hour by hour the journey of the fertilized ovum.

In the end, it was Miriam's other project that brought them one of the most exciting specimens. This was "the ova study," the quest to unite sperm and egg outside the body. After she succeeded in 1944, Hertig hand-delivered her specimens by plane to Baltimore, where they were examined and catalogued as Carnegie specimens 8260 and 8500.1. Frozen in division at just eight or nine days old, Miriam's achievement now represented the youngest specimens at Carnegie.

Today Carnegie specimens 8260 and 8500.1 can be found in the low, cream-colored storage warehouse across the street from the National Museum of Health and Medicine. They are nestled within

---

* This was perhaps less controversial than it sounds; in the 1940s it was not as common to believe that life began at fertilization. Another former lab assistant of Rock's, for instance, once wrote that in aiding his research, she had witnessed not the beginning and end of human life, but "the 3rd stage cellular reaction by division of human ova which had been activated outside of the body by laboratory procedure."

the moveable white metal bookshelves, which also contain 1920s-era plaster-of-Paris models of embryos and slices of human brain preserved in glass. Most specimens here are named after the doctor who procured them—for example the Miller ovum or the Rock-Hertig ovum. But the label on Miriam's reads only "Proctor ovum 1." Nor does the official cataloging form mention her name. It only reads: "Sectioned egg presented to Carnegie Lab by Dr. Rock," and lists Rock as the donor.

"This is pretty much as early as it gets," says Elizabeth Lockett, the collection manager of the developmental anatomy collection, taking the glass slides out of their wooden box and placing them on the stage of the microscope.

Lockett peers through the microscope and adjusts the slide beneath the lens. "Where the heck is the smear on this thing?" she says. The machine whirrs, focusing in at 4x magnification, then 10x. Compared to the instrument she's using to view it, the specimen itself is ancient. The medium used to fix it has grown yellow with age, its label browned and peeling. Finally, the globular pink mass comes into view: a two-celled egg, like two fried eggs joined at the middle. Stained pink and purple, the cells were caught in the act of dividing, sandwiched within the zona pellucida. This is the earliest stage of the fertilized egg, known as Carnegie Stage 1. The specimen has seen better days. There is detritus on the glass. The color has faded a bit. The innards of the cells stay blurry, as if you're peering through glasses with the wrong prescription.

But in 1944, this slide represented a feat never before achieved by science—an achievement that made the great minds at Carnegie gasp with delight. "We are quite excited over it and are keen to see the sections," one Carnegie scientist wrote to Rock on July 12, 1944, after Miriam had sent him photographs and a detailed description of the embryos. They were the brainchild of Rock and Hertig, the scientific minds behind the project. But little recognition was given to the woman who brought them into being: a smart, tenacious researcher named Miriam Menkin.

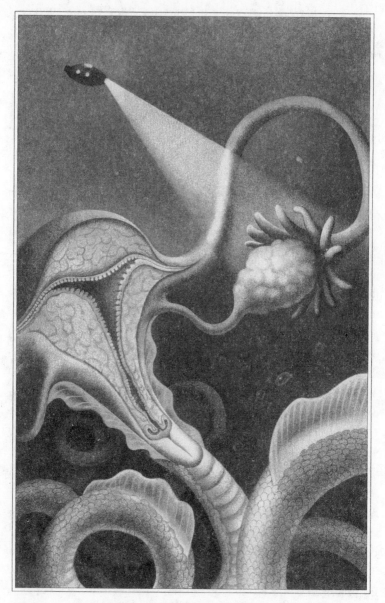

Not known, because not looked for
But heard, half-heard, in the stillness
Between two waves of the sea.

—T. S. ELIOT, "LITTLE GIDDING"

CHAPTER 6

Power

(OVARIES)

Jon Tilly didn't set out to rewrite the ovary. He just wanted to understand how ovaries age and, eventually, stop working. It was the early 2000s, and Tilly was an up-and-coming reproductive scientist at the new women's health center at Boston's Massachusetts General Hospital, where he ran a Harvard-associated lab that studied how germ cells in the ovary die off.

Ovaries, he knew, have one of the toughest jobs in the body. They start out pink and smooth, around the size of two cherry tomatoes. At puberty, a surge of hormones triggers the dormant eggs within them to sleepily awaken. From then on they're consistent workhorses, swelling to the size of two kumquats and coordinating a factory of eggs in every stage of being: waiting, growing, bursting, dying. Each time they jettison an egg, they have to heal over the crater and begin again, earning a new scar. Like the heart, they toughen up.

By menopause, most of that dynamic activity ceases. That's why, for Tilly, these two almond-shaped glands are windows into a woman's overall health—paired canaries in the coal mine for female aging. "Why do we age?" Tilly asks. "For women, aging is intimately associated with the ovaries. And unlike most other tissues of the body, the ovaries age long before anything else."

Understanding how ovaries age might hold the key to slowing down that process. If you could lengthen ovarian life span, women might be able to lead healthier lives after undergoing radiation and other ovary-blasting cancer treatments. They might even be able to stave off some of the problems associated with menopause, including osteoporosis, heart disease, and dementia. To understand this process, Tilly's lab zoomed in on the egg follicle, the functional unit of the ovary, and what factors lead to its eventual demise. Egg cells mature and die throughout life, but Tilly had discovered that eggs don't wither away at random. Their death is genetically pre-programmed: Each follicle contains within it a genetic blueprint for its own destruction.

"My lab was death," he says.

If you could hack into that program, he thought, you might be able to slow down the aging process by keeping eggs—and therefore, ovaries—limping along a little longer.

As he started running experiments, he started getting results that were . . . off. In one, his postdoctoral fellow, Dr. Tomoko Kaneko-Tarui, was treating female mice with chemotherapy drugs to kill off their egg cells. When she examined their ovaries afterward, she found that the number of eggs had dropped precipitously. But then, they started to climb back up. That shouldn't have been possible. She repeated the experiment again with her fellow postdoc, Dr. Joshua Johnson, and got the same result. The pair consulted with the rest of the lab to see if they had made a technical mistake. "We just kept getting screwy outcomes," Tilly says.

At the time, Tilly imagined the ovary as an hourglass, with each follicle as a single grain of sand. Starting before birth, these follicles begin trickling away, in the endless march toward death. Twenty weeks into gestation, an infant floating in the womb is revved up and ready to go, her ovaries bursting with as many eggs as she will ever have—6 to 7 million.* More than three-quarters die off before she's born. Each year until puberty, around 120,000 more bite the dust. By the time she has her first period, only 300,000 to 400,000 remain. Between the ages of forty-five and fifty-five, the majority of women run out of eggs completely. Their ovaries shut down, taking most of their essential hormones with them. The hourglass runs out. That's menopause.

To understand egg-cell death rate, Tilly's lab started counting how many eggs died over a certain period of time and comparing them to how many were left in the ovaries. Essentially, they were ovarian accountants. But the rate of egg death and the time that ovaries failed didn't add up. In adult mice, egg cells were indeed dying off constantly—but at the remarkable rate of one-third of the total oocyte pool every few days. If you extrapolated that rate out to humans, a woman's eggs would be gone by the time she hit twenty-five or thirty. For six months, they reran the experiments using different protocols, trying to figure out what went wrong. Nothing changed.

"You can't argue the numbers," Tilly says. "Follicles don't just magically reappear. There's only one way new follicles can show up, and that's to make them." Something was untwisting the cap of the hourglass and dropping more sand in. And Tilly had a hunch he knew what it was: stem cells. These pluripotent cells, which scientists already knew existed in bone marrow, blood vessels, and the testes, have the ability to give rise to specialized adult cells like bone cells and sperm. Could there be stem cells making new germ cells—aka eggs—in the ovaries as well?

When he went back to the literature, he realized that not all animals seem to have the system of gradual degeneration we accept in women. Female flies, birds, and fish create new eggs nearly every day. Meanwhile, almost all male animals produce new sperm; men produce more than a thousand sperm every time they *breathe*. From an evolutionary

---

* That means that, at one point, her mother carried not only the seeds of her daughter, but also her grandchildren—three generations in one body.

perspective, that made good sense: you don't want your precious germ cells sitting around for decades, growing stale and increasing the chances of genetic mutations. "It's silliness. It's just honest-to-God silliness," Tilly says. "Why would nature deviate only in mammals and say females only get a fixed pool of eggs at birth? If you just think about it from a logic perspective, that makes no sense."

If he was right, it would change everything science knew about how ovaries age. Maybe what you started with wasn't necessarily what you ended up with. Maybe women weren't simply running out of eggs. Maybe it was that their stem cells were too damaged to produce more—and those stem cells could be convinced to get back to work. "Before, we thought menopause was like death and taxes: It's coming. Nothing you can do about it," Tilly says. "But now it looks like that isn't the case."

He compares the idea of ovaries making new eggs to the idea that the brain generates new neurons—a finding the field of neuroscience considered ludicrous before the 1990s but is now accepted as scientific fact.

"It never occurred to me that it was remotely possible," he says. "Until we did that one experiment, and everything changed."

———

Tilly wasn't exactly destined to bring about an ovarian revolution. His path into science had started a decade earlier, with him standing chest-deep in a brackish, frozen river in the dead of winter. It was 1984, and he had just completed his undergrad degree, making him the first in his working-class family to finish college. Afterward, he returned to his hometown on the Jersey Shore and started building docks along the river as a construction worker, just like his two older brothers. That winter the ice had pulled the wooden pilings up from the docks, and his task was to cut a hole in the ice with a chainsaw, wade in, and put them back in place.

Standing there in subzero wind-chill temperatures, snow whipping around his face, he realized: This was miserable. "You know what? I just don't want to do this when I'm fifty, sixty years old," he recalls thinking. He had to go back to school.

That was easier said than done. Tilly was more than a year out of undergrad and had never been in a lab. He didn't look like your typical academic: he was big and burly, with unruly brown hair past his shoulders and a diamond stud in one ear. When he went back to his alma mater, Rutgers University, and begged every adviser possible to take him as a master's student, only one agreed. He was a reproductive biologist named Dr. Alan Johnson, who happened to work on chicken ovaries.* "I didn't pick him because of my interest in reproduction," Tilly says. "I picked him because he was the only faculty member who would take me."

That serendipitous pairing would set the arc of his career. Within a year he had published his first academic paper, a technical description of an enzyme in the chicken ovary that contributed to both ovulation and egg-cell death. When he saw his name in the journal, the hair on his arms stood up. "That rush, the rush of knowledge, hit me like running into a brick wall," he recalls. For the first time, he knew what he wanted to do with his life. But he didn't know how to keep doing it. When his adviser told him that the next step was to get his doctoral degree, Tilly was stunned. He'd never even heard of a PhD. He thought doctoral degrees only meant MDs. Here, he realized, was another way: "Boom: everything fell into place."

Ten years later, Tilly was in Boston directing the women's health center at Mass Gen, where he worked with cancer patients and infertility patients, reproductive biologists and IVF doctors. He had recently found that cancer therapies accelerated the death of egg cells, often leading to early menopause and the end of fertility. The drugs and radiation didn't just kill ovarian tissue—they hijacked the existing

---

* "Chickens? Are you kidding me?!" Tilly remembers thinking.

genetic program of cell death, speeding up the process. If you could identify that program, you could potentially protect those egg cells. He switched tack, devoting his lab to figuring out how to slow down cell death in the ovary.

Then, at age thirty-eight, Tilly was diagnosed with testicular cancer. Up until that point he had been laser-focused on his academic career, publishing papers, getting NIH funding, and working eighty-hour weeks to run his health center and his lab. He barely saw his wife and four-year-old son at home, because he was so busy trying to help other people get pregnant at work. When he was diagnosed, his oncologist told him to come in the next day for surgery, followed by radiation therapy. But thanks to his research, Tilly knew exactly what radiation did to fertility. He also knew that he eventually wanted more kids.

He put off his oncologist and froze his sperm, which he would later use to father two more children, both girls. Then he started asking female cancer patients at the hospital what they hoped to get out of their treatment. Women, he realized, had none of the options men did. "For men it's easy, you freeze a bunch of sperm and you don't worry about it," he says. "Women can't just freeze a million eggs." In 2000, egg freezing for future IVF was still in development, and many cancer patients didn't have the time or money to go through rounds of hormones that would cause them to ovulate a few extra mature eggs.* For the average cancer patient, there was virtually no counseling about their fertility after cancer.

Tilly started thinking about the bigger ramifications of his research. By the time he stumbled across the possibility of stem cells making new eggs in the ovaries, he was already committed to disrupting the way we treated women for cancer. Imagine, he thought, if you could make eggs on demand. Imagine using stem cells to reinvigorate the ovaries after chemotherapy. By harnessing the regenerative power of ovarian stem

---

* Today, advances in reproductive technology have made it possible to "ripen" immature human eggs for in vitro fertilization, or to freeze for later use.

cells, women might have new options when it came to having children and living healthy lives after cancer.

Again, he was struck with that tingling feeling—that he had discovered something no one else in the world knew. He saw himself as an explorer, charting new horizons: "We're rewriting the map. And the map has to be accurate."

———

The first hint that Tilly's big reveal was not going to go well came at a conference in March 2004. He stepped up to the podium to present his finding that mouse ovaries produced new eggs, using stem cells located near the surface of the ovary. Then he explained how his team had verified this astonishing claim. In one experiment, they identified meiosis—a specialized form of cell division—happening within the mouse ovary. In another, they genetically altered mice to carry a jellyfish gene that made them glow fluorescent green under UV light, and transplanted fragments of normal ovaries into them. Some of the green stem cells migrated into the new ovaries, producing glowing-green eggs enveloped within non-green follicles—proof the eggs had arisen from stem cells.

The talk, he remembers, was met with dead silence. When he returned to his seat, a colleague turned to him and said: "The field is going to want to pick this apart."

When the paper came out in the journal *Nature* a few days later, the media reaction was explosive. "Scientists Find a Way to Beat the Menopause," one headline read; "Breakthrough Turns Fertility Wisdom on End," read another. Unlike most scientists, Tilly was happy to speculate on what this finding meant for the future of female health and fertility. Maybe, he suggested to reporters, human ovaries were doing the same thing. Maybe ovaries weren't faulty organs that aged faster than any other and eventually wore themselves out—they were something more resilient, with the possibility of growth and renewal.

If so, "women could essentially grow back their ovaries after [chemo] therapy," he told *New Scientist*. "The possibilities are almost too numerous to mention."

Clinicians waxed breathless about the implications for female fertility. Dr. Marian Damewood, then the president of the American Society for Reproductive Medicine, said Tilly's work "could be the most significant advance in reproductive medicine since the advent of IVF." Dr. Roger Gosden, scientific director of the Jones Institute for Reproductive Medicine, agreed. "The ability to make more eggs would be a revolution in women's health"—as big a revolution as cloning Dolly the sheep was for embryology. Private philanthropists began approaching Tilly, offering him funding to focus on extending fertility.

Leaders in his own field weren't so quick to accept that the past half-century of ovarian science had been wrong. Other researchers tried to replicate his study and couldn't; some even accused Tilly and his team of shoddy science and falsifying data. One of his most outspoken critics was Evelyn Telfer, an oocyte researcher at the University of Edinburgh who had been studying ovaries since the 1980s. "The presence of such cells in other mammalian species remain to be proven," Telfer wrote in a rebuttal article. "If the dogma is to be debunked and a new one accepted, at least in the mouse, it is sure to be challenged and tested, as should all dogma. This is the way of science and scientists."

Others protested not the science itself but its delivery. Tilly had promised too much, too fast, said Dr. David Albertini, the oocyte expert who remembered Miriam Menkin from grad school. After the finding was announced, Albertini, then at Kansas University Medical Center, found himself being approached by women at fertility-preservation conferences who had learned about the possibility and wanted to use it themselves. He wished the first announcement had been more "subtle," both in the literature and in the media. "Phones were ringing off the hook with patients calling clinics long before the scientific community had a chance to evaluate that work," he told the journal *PLoS Biology.*

Much of the resistance against the idea, Telfer would later admit, was unscientific. Tilly was brash and confident, willing to make bold claims. He used phrases like "100 percent confident" and "unequivocal." "He's a go-getter, he's a Jersey guy," says Johnson, Tilly's old PhD adviser. To many, he seemed like a salesman, not a scientist—"evangelical, almost," says Telfer. "Jonathan is very good at putting his points across and defending his work very forcibly, and he has a right to do so," she said in 2007. "But he can upset some people because he's presented himself as a misunderstood visionary and that everyone else just doesn't 'get it.'"

At this point in his career, Tilly could take the heat. He had run a successful cell-death lab for years, had just been promoted from assistant to associate professor within Harvard Medical School, and had several grants from the National Institutes of Health. His younger colleagues weren't so lucky. Joshua Johnson, the first author on the 2004 paper, took lasting damage. After the paper came out, he remembers, colleagues in the field would flat-out avoid him at meetings. He was "utterly blasted" as a scientist. He stopped getting grants. He ultimately had to distance himself from stem cells and Tilly's lab, and broaden his focus out to ovarian function as a whole, just to get a job.

Looking back, Tilly sees his field's resistance to the idea of the female body making new eggs as running deeper than just an attachment to old ideas. It was about what the female body could and could not do. "The field was so entrenched in the idea that the ovary had no regenerative capacity, period," he says. "It came out of decades of the unfailing belief that the ovary was different from the testis, and that the rate of loss of oocytes from the ovary and eventual menopause in women was proof positive there was no regeneration. Because if there was, ovaries shouldn't fail." Women, it was thought, were designed to fail; the flaw in the machine was what made them women. It was in every textbook, every study, every data set.

Tilly had dared to question how women's bodies worked. To introduce the possibility of their lives looking different—giving birth later, staying healthier longer. For that, he believes, he was pilloried. "It was so nasty," he says. "It was personal. Questioning my integrity, did we lie, how dare I do this." To convince a stubborn field, he would need a smoking gun: definitive proof that ovarian stem cells existed in humans. That meant finding a cell within the female body that was unknown to biology—a white whale, an unknown shore.

Then Dr. Dori Woods came along.

———

Dori Woods's menopause journey started at the makeup counter. It was near the end of her PhD studies in ovarian biology at Notre Dame University in Indiana, and she was at the Clinique stand at the mall, looking for sweatproof makeup. She had recently started running and lifting weights, and her foundation kept melting off her face. As it turned out, the representative at the counter told her, Clinique had a whole line of waterproof products. Mostly, women bought them because they were going through menopause—"because of hot flashes," she added, fanning her face with her hands.

Up until that point, Woods hadn't spent much time thinking about menopause. Why would she? She was twenty-six and had her life ahead of her: she was recently married, about to start her professional career, with two kids planned for after she finished her PhD. With her round face, singsong voice, and athletic frame, people constantly remarked on how young she seemed. "That's when I really thought about it and realized: not just is this a huge industry from that perspective, but it's a major problem that impacts literally every facet of a woman's life, even down to the makeup that she's looking to buy," she says.

Woods left Clinique with a tube of waterproof mascara and a bottle of foundation. But the comment stayed with her. When she got home, she started doing research online. Menopause, she learned, was far more than just hot flashes. It was associated with an increased risk of stroke, heart disease, brittle bones, and even mental-health ailments like Alzheimer's and Parkinson's. One of the biggest studies on its health effects, the Study of Women's Health Across the Nation, followed more than three thousand premenopausal American women for ten years to find that the transition was associated with cardiovascular risks and permanent changes in blood vessels.

"That was a real eye-opener for me," says Woods. As an ovarian researcher, she realized, she might be able to do something about it.

For most people who study ovaries, the egg is the thing. It is the ovary's raison d'être, the main event, the cell from which all life arises. The egg, as Miriam Menkin saw it, is an entire world in a speck. That isn't how Woods viewed it. Today, as a forty-two-year-old ovarian biologist and director of biology graduate studies at Northeastern University in Boston, she considers the egg, well, kind of a waste. "It's an awful lot of work when you just use a couple of them—in my case two," she says, referring to her two daughters, Allison, age fourteen, and Marin, age eleven. What about the other four to five hundred she will ultimately ovulate in her lifetime? What about the millions of immature eggs that shrivel and die, many while a woman is still in her mother's womb, never to see the light of day?

"I mean, it's definitely important, and I really love those two cells and what they ended up becoming," she says, "but it's really the continued production of hormones and the drawbacks of having menopause that make me very interested in the ovary as a whole."

Woods started out enamored with the egg. When she started graduate school at Notre Dame, she ended up in the same lab as Tilly had been in seventeen years earlier, working under Alan Johnson, studying the chicken ovary.

For reproductive scientists, the chicken is an ideal study animal. It has only one ovary (like most birds, it's born with two, but the right one withers away soon after birth). The one that remains is massive, big enough to palm in your hand, and looks like a stalk brimming with fat, shiny, yellow tomatoes. Each one is an egg follicle, which will eventually become the chicken yolk. Within one bunch, the largest egg is designated next in line for the throne. Behind it is an endless succession of slightly smaller eggs patiently waiting their turn. This setup allows researchers to study eggs in every form of development, from tiny primordial specks to the massive, bulging yellow yolks preparing to travel down the one massive oviduct.

Woods wanted to know one of the ovary's most enduring mysteries: what force selects certain follicles to be ovulated and others to waste away? The more she looked, the more she realized that the secret lay not in the oocyte but in another type of cell, just a nanometer to the side: the granulosa cell. These were the cells that produced hormones and nurtured the developing egg. While most people saw the egg as the conductor of the symphony, to her, granulosas were the ones making the ultimate decision of which eggs lived, which died, and which were ovulated. Through their hormonal signals, they were orchestrating egg selection.* Ovarian scientists usually "want to work with the thing that makes the baby," she says. "But really if you're going to look at the heart of the problem, it's in these granulosa cells."

Look at a follicle through the microscope, and you'll see a halo of tiny, translucent cells enclosing the egg like a bubble. These are granulosa cells, a word that comes from the Latin for granule, or small grain. They provide nutrients and chemical signals to the egg, ensuring that it eventually matures and ovulates. In the past, they've been defined more as supporting actors than characters in their own right. Research-

---

* And not just estrogen—recent research suggests that testosterone plays a key role in which follicles develop, and are ultimately recruited, each month.

ers refer to them as "nurse" cells or "companion" cells—handmaidens to the queen bee herself. IVF doctors don't much care for them; they regularly pluck them off and throw them out so they can get at the egg to fertilize it.

But in the 1990s, scientists began learning that both cells are necessary players in a complex web of intercellular communication. Each egg depends on its surrounding granulosas to help it develop: "they are essential for the oocyte's ability to exist and function," says Dr. John Eppig, a reproductive biologist who first described the crosstalk between egg cells and their partners. An ovary is not just a basket of eggs: it is a crackling network of communication, alive with signals that pass back and forth between and within follicles. Most of those signals are sent by the granulosas. Both cell types are active, dynamic partners in an exquisitely coordinated duet.

"One cell type can't function normally without the other," says Eppig. Rather than nurse cells, he calls granulosas the egg's "buddies."

The egg and granulosa each have tiny, snaking tentacles—called processes or microvilli—that reach across the divide to touch each other and communicate. "Like a hand grabbing onto a watermelon," says Albertini, who was one of the first scientists to identify these processes in the 1970s. The two send signals across this connection, not unlike the way a synapse transfers an electric charge across a junction between neurons. These tentacles connect granulosas to eggs, and granulosas to granulosas, creating an "incredibly hairy ball" of communication that resembles a brain, as Albertini puts it. These junctions are the most massive in any tissue, reflecting the complexity of the conversations going on.

Moreover, granulosa cells have a crucial second function: to pump out hormones to the brain and body. When a follicle ripens to a certain size, its granulosa cells multiply and swell to begin making estrogen, progesterone, and testosterone—chemical messengers that make their way to almost every tissue in the body. The hypothalamus gives the initial signal to the pituitary gland, which then tells the ovaries to start making the hormones estrogen and progesterone, which start the cycle. The ovaries, in turn, tell the brain to tamp down or continue the cycle. The gonads are thus nodes in a bigger feedback loop between brain and body.

For decades, scientists have defined menopause as the event that happens when a woman runs out of eggs. Only, it's not that simple. She isn't just running out of eggs but out of the hormone-pumping granulosa cells that surround them. When the last granulosa fizzles out, this brain-ovary feedback abruptly ends. It's as if an invisible string has been cut, one connecting her gonads to the rest of her body.

Few had thought to try to prevent this process, because of the assumption that there is no way for a woman to make more eggs. But Woods realized that if you could keep the ovary stocked with granulosas, you could keep that crucial feedback loop going for ten, twenty more years, potentially protecting a woman from some of those cascading health risks. Menopause—including early menopause triggered by cancer treatment or environmental factors like smoking—wasn't inevitable. If you understood exactly what was going on in that complex hormonal waltz, you could potentially disrupt it. "It's not about the transition," she says. "It's about extending the quality of life."

Two years later, she got a call from Jon Tilly.

Toward the end of her graduate studies, Woods was looking for a job. Her adviser, Alan Johnson, got her in contact with Tilly's lab. As it happened, Tilly was looking to hire a new reproductive scientist to expand his women's health center—but at that point, his lab was still a hotbed of controversy. So even though he'd put out feelers for her, Johnson warned Woods against taking the job. She was just starting out on her path into academia, and Tilly's lab was radioactive.

Woods hesitated. Unlike Tilly, she preferred to stay out of the limelight. But she was intrigued enough to fly to Boston for an interview.

Once she got there, her doubts subsided. The prospect of working in a hospital and having major clinical impact appealed to her. It was a way to go beyond chickens and into humans. Plus, she told herself, she wasn't going to work on stem cells, the center of the controversy. She would be a junior faculty member in her own right, studying granulosa cells—the soma, the dance partners. "That was pretty naïve, huh?" she says now.

The first time the two met, both say, it was kismet. Tilly puts it in stronger terms: "It was fate and destiny." By that point Tilly was completely focused on renewing egg cells. He saw the egg as the nucleus of the follicle, and the key to his quest for extending ovarian function. Before Woods arrived, he had become "completely germ-cell centric," he says. "I had totally lost the desire to think about granulosa cells."

Woods, of course, had eyes only for granulosas. But both immediately realized the importance of the other. In one of their first conversations, Woods pointed out: if stem cells are growing new egg follicles, they have to be growing new granulosas too. They still hadn't settled the question of which ran out first, the egg cell or her partners.

Woods took the job, and she and Tilly settled on the ultimate goal of building an "artificial ovary": a collection of lab-grown cells that would deliver hormones and other crucial substances into a woman's body long after menopause. They imagined these ovaries would first be used for cancer survivors whose ovaries had been obliterated by chemo or radiation, women whose ovaries had been damaged by chronic alcohol use or smoking, and women with Turner syndrome, a genetic condition that stunts ovarian development. But there were more far-off possibilities too: an artificial ovary could one day be used as a nursery to mature eggs that have been altered using CRISPR editing technology to prevent genetic diseases, before reimplanting them back into a woman using traditional IVF.

"It's a possibility, and technically it's on the horizon," Dr. Woods says.

There was no reason an artificial ovary had to look like a biological one. Woods imagined it as flat and round, like a silver dollar. Nor would it have to be attached to the existing ovary. Endocrine glands can take root, grow blood vessels, and make hormones anywhere in the body. It might be inserted just under the skin in the arm, or leg.*

Still, Woods wasn't planning on touching those stem cells. She just wanted the granulosas. But granulosas, however important, couldn't survive without their organizing nucleus. The functional unit of the ovary is the follicle—the egg and its halo cells—just as the nephron is the functional unit of the kidney. The two share a fate: without an egg cell, the granulosas waste away. To study them together, she began differentiating pure stem cells into germ cells and granulosa cells in a petri dish—but the germ cells soon proliferated. "These pesky germ cells were screwing up my culture . . . they wrecked everything," she says.

Frustrated, she searched for a way to isolate the egg cells and rid them from her samples. In 2009, a Chinese team reported they had also successfully purified ovarian stem cells from mice, reigniting the stem-cell debate. Woods began trying to mimic their method to deplete the egg cells from her culture. Within months, she had succeeded.

When Tilly realized what she'd done, he asked if she could do the same thing in ovarian tissue. She was confident she could. But after nine months of failure, they were both ready to give up. Finally, she tried a slightly different method, called flow cytometry, and managed to purify out a few of the long-sought-after stem cells from the ovarian tissue of a mouse. They would eventually do the same in humans, cows, monkeys, salamanders, and even giraffes. Now Tilly had his smoking gun—and they both had a way to build new ovarian function.

Tilly and Woods immediately recognized that an ovary that renewed itself gave rise to another possibility. Both saw the ovary the same way:

---

* It's actually a better idea to keep it far away from the ovaries, Woods says, to prevent the possibility of an accidental, unwanted pregnancy.

not just as an egg factory but as the key to prolonging health. "There's a grander goal on the table," says Tilly. Thanks to the rise of IVF and other reproductive technologies, we have myriad solutions for different types of infertility. But delaying menopause? "I think that the benefits overall to half the population are really clear," Woods says. If you could extend ovarian life span, you could also extend a woman's healthy life.

—

Until the mid-seventeenth century, the ovaries had no name of their own. They were known only as "female testicles," a name that reflected the mistaken belief that they produced female sperm. Dutch anatomist Regnier de Graaf, namesake of the Graafian follicle, had the foresight to challenge this assumption. De Graaf argued that women were not just ornaments or "bodies without function" but were essential to reproduction. Their bodies had purpose. "Nature had her mind on the job when generating the female as well as when generating the male," he wrote.

After dissecting the ovaries of recently mated rabbits, de Graaf found what he believed were eggs: knobbly protuberances that dotted the ovary.* He thus adopted the name ovaries. "The common function of the female 'testicles' is to generate the eggs, foster them, and bring them to maturity," he wrote. "Thus, in women, they perform the same task as do the ovaries of birds. Hence they should be called women's 'ovaries' rather than 'testicles,' especially as they bear no similarity either in shape or content to the male testicle."

For centuries, science agreed with his assessment: the main purpose of the ovary was to produce eggs. "Ovary," after all, means "egg keeper." But that wasn't all they did. The first clues that ovaries did

* They were actually ruptured follicles; as we saw in Chapter 5, it would take until 1827 for scientists to directly observe the mammalian egg.

something more came from an unlikely source—what happened when surgeons took them out.

———

Twenty-three-year-old Julia Omberg of Rome, Georgia, had rarely had a normal period in her life. Her menstrual cycle brought with it violent convulsions, inflammation, and rectal bleeding, leaving her confined to her bed and nearly comatose. For years she used morphine to dull the pain, and at times she had given up the will to live. Her physician, Dr. Robert Battey, a Confederate surgeon during the Civil War, thought he knew the root of her suffering. "If I could but divest her of the source of her menstrual molimen—viz., the ovaries—there would be hope for her," he wrote. Battey tried everything to stop her "vicious and abnormal ovulation," including injecting her uterus with silver nitrate (used at the time to treat venereal diseases), but to no avail.

On August 17, 1872, he performed what was then a rare and dangerous surgery: the removal of both ovaries. In the days before widespread sterilization, such a procedure was usually a death sentence. Yet after ten days in which he housed her and sat by her bedside, Julia slowly began to recover. The most obvious consequence was one that surprised many of Battey's colleagues, and possibly Battey himself: she never had her period again. Though it was generally known that ovaries played a role in female sexual development, scientists had not yet linked them definitively to the menstrual cycle, which was still largely a mystery. Nobody knew how signals to menstruate moved through the body—through nerves or blood.

Battey realized that removing the ovaries would "arrest ovulation and produce the change of life," a polite way of saying it caused early menopause. He boasted that women who had been utterly incapacitated by painful menses could now be returned to health virtually overnight. One of his patients reportedly wrote that, after suffering pain so bad she was driven to suicide, "my life now seems a new one . . . I am now a well, happy, and cheerful girl, and do not feel like the same person at all." Ovarian removal would soon become a panacea not just for menstrual symptoms but for a variety of ailments—from masturbating to melancholia to epilepsy—together termed "menstrual madness."

After Julia's operation, Battey would remove the healthy ovaries of hundreds of women, many of them in their twenties and thirties. Those who survived—the mortality rate was close to 1 in 3—were effectively sterile. "Battey's Operation" gained widespread popularity in Europe and America, with one doctor estimating that 150,000 women had gone under the knife by 1906.* Hailed as a "triumph of surgery," ovariotomy (a word formed from the Latin for "ovary" and "cutting") became the first commonly performed abdominal surgery, paving the way for a renaissance of surgical advancements. "It was like the discovery of the Californian diggings or the African diamond fields," wrote the president of the Royal College of Surgeons in 1886. "The way was cleared for all prospectors, and the benefits spread world-wide."

Battey was hailed as a daring surgical pioneer who, like his colleague and advocate, gynecologist James Marion Sims, had laid the darkened female body and its secrets bare.† "It was Battey who invaded the hidden recesses of the female organism," as one admirer put it, "and snatched from its appointed place those delicate little glandular bodies whose mysterious and wonderful functions are of such high interest to the human race." As his surgery spread, Battey's confidence in deciding to cut quickly grew. "I decide, in such cases that the sacrifice of the

---

* A large number of them were in mental asylums, revealing surgeons' eugenic attitudes towards institutionalized women. Even if the patient died, "the surgeon can console himself with the thought that he has brought about a sterility in a woman who might otherwise have given birth to an insane progeny," claimed one promoter of the surgery.

† Sims, in fact, was the one who named the procedure after Battey and helped make it respectable among surgeons. Ultimately, his overuse of the operation—as well as his insistence on inviting dozens of spectators into the operating room to witness his surgeries—led to his forced resignation from the Women's Hospital of New York in 1874.

ovaries shall be made, and I believe I do God a service when I sacrifice them," he wrote in 1886.

Yet, as with clitoral amputation, doctors soon raised concerns about the enthusiasm their colleagues seemed to have for this particular procedure.* Although Battey claimed that "there is no loss of the womanly graces," some noted with concern that in women whose ovaries were removed, their voices deepened, their breasts flattened, and they grew an "unwomanly pilosity on the chin, upper lip, and chest." Critics referred to his surgery as "spaying," "unsexing," or "castration of women." Naturally, there are few records of what women had to say about it, besides a few glowing (and likely doctored) reviews that appeared in medical journals.

Though it would ultimately fall into disrepute, ovariotomy would leave a lasting scientific legacy. First, it cast new medical interest on the role of the ovaries. After Battey's first operation in 1872, a member of the Gynaecological Society of Boston suggested the society change its motto from the Latin "Propter Uterum" (short for "a woman is a woman on account of her uterus") to "Propter Ovarium"—thus continuing a time-honored tradition of trying to reduce women to one organ or substance. Secondly—and incidentally to his original aim—Battey helped advance the understanding of the ovaries' role in reproductive physiology. The results of his operations made it abundantly clear that ovaries were somehow conducting a woman's menstrual cycles.

"During the last twenty years perhaps no organ in the body has been so much written about as the ovary," wrote one eminent Scottish surgeon in 1883. "To the naked eye nothing could look more uninteresting and unimportant than a human ovary; and yet upon it the whole affairs of the world depend. As far as the individual owner of the gland is concerned—certainly for her comfort, and, if we take with it

---

* The "father" of that surgery, Isaac Baker Brown, also performed ovariotomies, including on his own sister.

its appendages, for her life as well, it is the most important organ in her body." Welcome to the age of the ovary.

———

Dr. Eugen Steinach came of age just as the study of glands and their secretions were coming into full force. The Austrian physiologist was in Prague, conducting research on the eyeball, when these "organ juices" first got their name: In 1905, physiologist Ernest Starling named them hormones, from the Greek verb "to excite, provoke, or arouse." What made them so special was their ability to travel throughout the body to their far-off destinations: they were chemical warheads on a specific mission.

Forget the soul, forget the brain. The human essence was glandular. The thyroid, the pancreas, the adrenal—these factories pumped out products like adrenaline, insulin, and cortisol that traveled through the bloodstream across the body to find their targets. They were all governed by one master gland, the pea-sized pair of lobes at the base of the brain known as the pituitary. By 1922 the *New York Times* would mockingly lament that "a war-ridden world has given place to a gland-ridden world." Glands were like DNA today: they represented the invisible forces within us that make us who we are, the secret behind vital life processes.

Steinach had seen firsthand the dramatic effects of removing glands from the body. Born in 1861 in a small town in the Austrian Alps, he grew up helping his grandfather raise cattle. He was struck by the consequences of castration, which "transformed the wild bullock into the patient ox, the lean pugnacious cock into the fat stolid capon, the tough hen into the tender poulard, and the fiery stallion into the amenable gelding," he wrote in his 1940 autobiography, *Sex and Life*. Later, as the director of the Physiological Section of the Institute of Experimental Biology in Vienna, Steinach became convinced that the key to sex traits lay in the ovaries and testicles—to him, "the very citadels of sex."

"Everyone knows even without books that men are generally hardier, more energetic, and more enterprising than women, and that women show a greater inclination for tenderness and devotion, and a love of security together with a practical aptitude for domestic problems," he wrote.* Ovarian and testicular hormones seemed to give a scientific reason as to why. In the 1890s, to test his theory, Steinach performed a series of experiments in which he removed testicles and ovaries from young rats and implanted them in the abdomen of a rodent of the opposite sex.† The glands took root, growing new blood vessels and drastically altering the appearance and sexual behavior of the animal.

In one case, Steinach took a rat pup, cut off his pea-sized testicles, and grafted another rat's ovaries into his stomach. He grew up stunted, but his breasts and nipples swelled, producing milk. As he reached adulthood, he showed what scientists consider "stereotypical female behavior": When another male tried to mate with him, he raised his hind feet and struck backward to wriggle away from the unwelcome suitor. Pups recognized him as female, following him around in pursuit of milk. He graciously nursed them—a maternal behavior that Steinach considered the highest expression of femininity. "The implantation of the gonad of the opposite sex," he concluded, "transforms the original sex of an animal."

Steinach didn't stop there. In the late nineteenth century, biologists had just discovered that female and male gonads—the ovaries and testes—developed out of the same embryonic structures in the womb. In other words, every human being started with the potential to develop into a female or male; each ovary is a testicle-that-wasn't, and each testicle a once-potential-ovary. Steinach believed that there was

---

* Wait, what?

† He was building on the experiments of hormone trailblazer Arnold Berthold, who in 1848 transplanted testicles into the intestines of castrated roosters and noted that they grew back their combs and started strutting again.

thus no true, 100 percent man or 100 percent woman. Each retained a shadow of the other inside them, "slight remnants of the stunted sex," as Freud, Steinach's contemporary, put it. What if those dormant parts could be reactivated?

In another set of experiments, Steinach neutered a litter of male mice, and implanted them with both ovaries and testes. These animals grew into what looked like large, powerful males—but with breasts and nipples that lactated. Most remarkably, they cycled between typically male and female behavior, attacking other males one month and offering themselves up for mounting the next. To him, these experiments proved that animals were not simply preprogrammed to develop into males or females. Glands and their hormones determined which direction they would go—and glands could be modified. By turning one knob and then another, you could optimize human health and transform the psyche.

The possibilities were endless: changing sexual behavior, "curing" a lack of masculine or feminine traits,* or even restoring youth. To Steinach, sex and aging were intimately intertwined. He thought of aging as essentially a process of losing one's sexual characteristics. "It has frequently been said that a man is as old as his blood vessels," he wrote. "One may have greater justification for saying that a man is as old as his endocrine glands." At the time, notorious gland peddlers were swallowing testicular slurries, selling dubious ovarian extracts, and grafting slices of goat and monkey testicles onto men with the same promise.

Steinach had a simpler, less horrifying idea. In his experiments with

---

* In a stunt that was controversial even in its time, Steinach once worked with urologists to attempt to "convert" homosexual men by castrating them and surgically attaching the testicular tissue of other men. The researchers concluded, with no evidence, that they had successfully changed the men's libido from homosexual to heterosexual.

rats, he homed in on cells in testes that produced hormones.* When
he cut the spermatic duct in male rats, it seemed to trigger an increase
in these hormone-producing cells, and a subsequent increase in sexual activity. Lethargic, skinny, listless rats gained weight, developed
glossy fur, and became sexual athletes. Some decrepit animals now had
intercourse as many as nineteen times per day. The same thing, he rea-
soned, could restore youthfulness in aging men. He developed a simple
operation—essentially a one-sided vasectomy—that he believed would
cause sperm-producing cells to shrivel up and the hormone-producing
ones to proliferate, flooding the bloodstream with manly hormones
and initiating a tidal wave of new energy and vitality.

Snip-snip, and it's done! Just like that, the glow of youth is yours
again. Lower your blood pressure, sharpen your eyesight, restore the
ability to remember your grandchildren's names. The strength, the
energy—and yes, if you must ask, the sexual prowess—of yesteryear is
at your fingertips.

By the 1920s, his procedure, known as the Steinach operation or
simply "getting Steinached," had exploded in popularity. The poet
William Butler Yeats did it at age sixty-nine and couldn't stop rav-
ing about his "second puberty." "It revived my creative power," Yeats
wrote in 1937, before penning a collection of poems ranked among his
best work. "It revived also sexual desire; and that in all likelihood will
last me until I die." Freud did it, too, at age sixty-seven. He hoped it
would improve his "sexuality, his general condition and his capacity for
work"—as well as prevent his jaw cancer from coming back. (It didn't.)
"Steinaching" rocketed in popularity among wealthy, aging men who
wanted to turn back the clock. It was the Viagra of the 1920s—or,
more aptly, a forerunner of modern hormone replacement therapy.

---

* Like the ovaries, the testes have two types of cells: ones that produce hormones
(known as Leydig or interstitial cells), and ones that produce fledgling sperm (known
as Sertoli cells).

Don't worry, ladies, there was a version for you too. The female equivalent was a shot of low-dose X-rays to the ovaries. Similarly, it promised to kill off germ cells and lead to the proliferation of hormone-producing cells instead. Ovary-zapping and procedures like it—some of which involved grafting part of a monkey ovary onto a woman—promised to "convert grandmothers into debutantes." Perhaps Steinach's most famous patient was the American novelist Gertrude Atherton, who had the X-ray treatment in her mid-sixties during a bout of writer's block. Afterward, she "had the abrupt sensation of a black cloud lifting from my brain, hovering for a moment, rolling away. Torpor vanished. My brain seemed sparkling with light." In 1923, she published the best-selling novel *Black Oxen*, about an aging European countess who has the same procedure and is returned to her former youth and beauty.*

Skepticism of Steinach's claims abounded, with many doctors sarcastically dubbing it the "Steinach miracle." Yet the public hype around Steinaching revealed a general faith in the power of science for progress. Atherton, in her lifetime, had seen the rise of germ theory and the scourge of epidemics wiped off the planet. In the wake of World War I, she recommended that Germany "Steinach" its population to reactivate the elderly and create a new breed of supermen to recapture the nation's lost glory. One of Steinach's scientific contemporaries wrote that "rejuvenation, by restoring mental faculties together with the love and ability to work, endows life anew with high, ethical riches."

Sex glands represented the promise of a brighter future—the ability to restore youth, vitality, and meaning to a weary yet hopeful generation.

---

* After its publication, another New York surgeon would beg her to let him try an ovary graft on her, asking her not to "blame a couple of innocent inoffensive sheep glands that are at heart trying their best to help you."

———

Before long, medicine had replaced glands with the hormones they produced: in the case of the ovaries, estrogen. It would take four tons of pig ovaries and untold gallons of human urine (kindly donated by pregnant friends), but in the 1920s, American scientists Edward Doisy and Edgar Allen would finally isolate this substance from humans. Although it had multiple effects, they named it for one conspicuous effect it had in rats: it produced estrus, the period in which a female is fertile and sexually receptive.* The word "estrogen" comes from the Latin (originally Greek) *oestrus*, "gadfly," and suggests a mad, frenzied passion as if induced by a buzzing insect. In fact, humans don't enter estrus; women are sexually receptive whenever they feel like it, and ovulate about once a month regardless. Yet estrogen would soon become known as distilled femaleness, the essence of what it means to be a woman.

One of its first uses was for women going through menopause. Steinach, for one, was careful to note that menopause was not a disease but a natural phase of life. "But is it really essential," he added, that women "should endure hot flushes, heat waves, dizziness, palpitation, ringing in the ears, depression, hysterical crying, anxiety, sleeplessness, itching, pain in the joints, irritability, and other tormenting complaints?" Starting in 1923, he worked with a German pharmaceutical company to help develop one of the first oral estrogens, Progynon B. Marketed as the "female cycle hormone," it was called by one admirer "the very essence of Eve."

In 1930, Steinach mailed a small, hexagonal glass bottle containing

---

* To determine this, they injected the substance into mice whose ovaries had been removed and examined their vaginal secretions under the microscope for signs of growth—a method recently developed by Greek gynecologist Dr. George Papanicolaou. Papanicolaou would later refine his test for use in humans as a tool for picking up early signs of cervical cancer; today it is known as the Pap smear.

several tablets across the pond to Atherton, his old patient. She would take Progynon daily until the day she died, although she reported that "when one has reached the venerable age of ninety, one cannot expect too much."

Steinach was further probing estrogen and the feedback between the ovaries and the pituitary gland when World War II broke out. In 1938, while he was on a lecture circuit in Switzerland, Germans overtook Austria and burned his institute and archives to the ground.* He and his wife, Antonia, both Jewish, fled to Zurich, where she died by suicide soon after. He died years later, disillusioned and alone, in 1944. But the effort to understand the essence of the so-called "female hormone" continued. Estrogen became a boon for pharma, leading eventually to the development of the birth-control pill and modern hormone therapy for menopause. By adding estrogen (and later, progesterone) into the body, companies promised to alleviate all the problems associated with menopause—a convenient way to market their product to virtually all women.

The trick was, pharmaceutical companies first had to sell menopause as a disease, with estrogen as the cure. And they did—with gusto. In the 1950s, estrogen cream was marketed to help women over thirty-five look younger and avoid wrinkles. Using it, "you have a new assurance. A new poise. Your husband looks at you with new interest," an ad from 1950 promised. It was also marketed to the husbands themselves. "It is no easy thing for a man to take the stings and barbs of business life, then to come home to the turmoil of a woman 'going through the change

---

* The same fate (or worse) would likely have befallen Freud, but for the quick action and patronage of his favorite princess. That year the Gestapo raided his publishing house, searched his apartment, and briefly arrested his daughter Anna. Marie helped arrange for Freud—now 82 and still plagued by jaw cancer—and members of his family to leave the country and settle in London. She also ensured that his library, antiquities, and iconic couch swiftly followed suit. He enjoyed a little over a year in that country before dying of cancer on September 23, 1939; his ashes were placed in an ancient Greek vase, a gift from Marie.

of life,'" claimed an ad from the '60s. But give her estrogen, and "she is a happy woman again—something for which husbands are grateful."*

The peak of this trend was the best-selling 1966 book *Feminine Forever*, which bleakly—and inaccurately—portrayed menopause as the loss of womanhood, youth, and mental health. "The unpalatable truth must be faced that all postmenopausal women are castrates," opined its author, a New York gynecologist named Robert A. Wilson, in a medical journal in 1963. "A man remains a man until the very end. The situation with a woman is very different. Her ovaries become inadequate relatively early in life." He argued that menopause was a disease of deficiency, similar to diabetes or thyroid disease. Note that Wilson was paid by the companies that made hormone therapy treatments, which may have (read: definitely) biased his conclusions.

Menopause wasn't the only use for estrogen. In 1940, Steinach had recommended it for maladies including painful periods, a lack of periods, infertility, sterility, frigidity, baldness, migraines, and general PMS symptoms—a list nearly as long and varied as the indications for ovarian removal had been in Battey's time. By restoring women's well-being during the tumultuous time before menopause, he wrote, it could even help keep imploding marriages intact.

Estrogen's uses broadened out further still. At places like Johns Hopkins University, it was used to treat children born with intersex conditions, encouraging breast growth and pushing a patient toward a more "acceptable feminine appearance." It was also used as "chemical castration" for those who exhibited deviant sexuality; a noted victim was Alan Turing, the founder of modern computer science, who was arrested for homosexuality in 1952 and forced to take estrogen pills. The pills made him impotent and depressed, and caused breast growth. Two years after starting them, he killed himself. Starting in the 1920s, some doctors also discreetly used the hormone to help transgender

---

* This ad was for Premarin, a form of estrogen made from the urine of pregnant ("pre") mares ("marin") that is still widely used today.

patients transition from male to female; a few patients were daring enough to get it under the table.

———

All these uses went by the same logic: that estrogen was a feminizing substance, used whenever you needed to add femaleness or subtract maleness. In reality, of course, estrogen is far more than a sex hormone. It contributes to the growth and development of all bodies, helping promote brain development, maintain heart health, regulate lipids, enhance insulin sensitivity, lower glucose, and normalize liver function. It is particularly important for bone density and closure; men with a rare inability to process estrogen end up growing taller and taller, their bones failing to knit together. That's why, in her 2000 book *Sexing the Body,* biologist and gender scholar Anne Fausto-Sterling suggested that the term "sex hormone" be changed to "growth hormone," to reflect the fact that these substances affect nearly every cell in the body.

Nor is estrogen merely the feminine equivalent—and antagonist—of testosterone. Far from polar opposites, the two are part of the same developmental chain: enzymes in the body convert testosterone into estrogen, meaning that in any body where you find one, you'll usually find the other. Rather than canceling each other out, they often work as a team to influence reproductive health across both sexes. In women, cells in the ovaries secrete small amounts of testosterone; some is converted into estrogen, while some remains as testosterone and contributes to ovarian health, bone health, and mood and libido. In men, estrogen produced by the testes and adrenal glands is crucial to the growth and development of sperm, brain development, and mood and libido.

Thinking of estrogen (or testosterone) as just a sex hormone actively blocks us from exploring its other myriad, important effects, as scholars Katrina Karkazis and Rebecca Jordan-Young write in their 2019 book

Testosterone: An Unauthorized Biography. "It's a transcendent, multi-purpose hormone that has been adapted for a huge array of uses in virtually all bodies," they write of testosterone. Estrogen, too, is a diffuse, dynamic, ever-changing substance, one whose effects transform depending on its environment, the other hormones it tangos with, and the cues it gets from the brain and body. It transcends sex, and defies its name.

In the decade after the release of *Feminine Forever*, estrogen sales in the United States quadrupled. By the 1970s, standard hormone therapy was touted as a panacea for all the ailments of menopause. In 1975, estrogen was the fifth most prescribed drug in the United States. But hormone therapy for menopause was really more of a stopgap, say Woods and Tilly. Certainly women facing brittle bones, hot flashes, night sweats, and vaginal dryness could benefit. But later studies would find hormone therapy did not necessarily protect women from long-term problems like heart disease and dementia, and could increase the risk of stroke and breast cancer.

"To market hormone therapy as a replacement for ovary function was just wrong—scientifically, biologically, in every manner," says Tilly. "Hormone therapy, since day one, was destined to fail." He and Woods want to add another tool to the toolbox—one they argue is longer-lasting, more natural, and more dynamic.

———

Yet, even if the science is solid, the question remains: is implanting artificial ovaries really something we want to do? While restoring the menstrual cycle in cancer patients is fairly noncontroversial, the idea that all women would want to delay or prevent menopause is by no means settled.

To say that the ovaries "fail" and need to be revived is a radical claim that lacks evidence, says Dr. Jen Gunter, a San Francisco–based gynecologist and author of the 2021 book *The Menopause Manifesto*. To her, menopause avoidance as a "cure" continues to treat menopause as a disease rather than a normal phase of life. She sees it as part of a larger trend of pathologizing women's bodies and the natural aging process—the modern version of *Feminine Forever*. (After all, men go through similar age-related fertility and health declines, yet we don't say they are in "erectopause.") Before experimenting on real women, Gunter believes we need to examine the biases that may be lurking in the science.

Yes, it's true that the brain-ovary feedback loop essentially stops with menopause.* But why do we think it should be reattached in the first place? "A fifty-four-year-old ovary should not be making estrogen. It's not supposed to be," she says. "So I think it's pretty bold to say that we should have ovarian function until the age of eighty." Some consequences of menopause may indeed be driven by these brain changes. But given that we don't fully understand them yet, we can't know if restoring them would be beneficial.

Menopause itself, Gunter stresses, is not the problem. It's the unwanted consequences that can come with it—and those vary for every woman. That's why the first question she asks her own patients approaching menopause is: What, exactly, is bothering you? For some it may be hot flashes and brain fog; for others it may be a family history of heart disease, osteoporosis, or dementia. Whatever it is, she needs to pinpoint the issues before she can address them. Similarly, researchers pursuing experimental treatments need to ask exactly what condition artificial ovaries are meant to treat, she says: "And if you tell me the problem is menopause, I'm going to tell you you're a misogynist."

When Gunter, who is 55, hit menopause, she opted for estrogen therapy. "Why would someone like me want to take this therapy with unknown long-term consequences to extend my ovarian function?" she says. "I have a medication that has fifty years of data on it." We have far less evidence about the effects of having working ovaries in women who

---

* That doesn't mean the ovaries are sitting around doing nothing. Even after menopause, they pump out trace amounts of estrogen, testosterone, and other hormones.

are sixty-five or seventy. For one, we don't know the consequences of restarting the cycle of growth and shedding of the uterine lining in older women.* For another, more research is needed to ensure that implanting hormone-producing tissue doesn't trigger reproductive cancer in the same way that exposure to high levels of estrogen has been linked to breast cancer.

Despite these concerns, some doctors are already forging ahead with the anti-menopause goal. In the 1990s, Dr. Sherman Silber, director of the Infertility Center of St. Louis, developed a strategy for cancer patients who later may want to become pregnant: he freezes tiny strips of their ovaries before chemotherapy, then attaches them back afterward, one by one. (He usually attaches the tissue directly to the ovary so that eggs can reach the Fallopian tubes—a technique that has led to the birth of more than eighty babies.) He's since moved on to doing the same for women who want to delay childbirth for career reasons, or who want to push back menopause. "We can beat the biologic clock of the ovary," Silber states. "We're forced to do it for cancer, but we can really do it for anyone now."

An IVF doctor in the UK has also begun freezing strips of younger women's ovaries, with the promise of reimplanting them down the line to stave off menopause. As of 2020, at least eleven women had paid more than $8,000 each to begin the process.

To Dr. Evelyn Telfer, the ovarian researcher from Scotland, these kinds of interventions are premature. "Is that really a good thing to do? Do we know enough information about that? Personally I would say we absolutely don't," she says. After all, she adds, "if you've gone through puberty it can be difficult as well. But we would never advocate halting puberty."† However, Telfer does believe that there is huge room for

---

* Yes, this means that artificial ovaries would most likely restart the menstrual cycle—a major drawback for many women.

† Actually, some physicians would. Puberty blockers are an emerging treatment for children who begin puberty too early, and for gender-nonconforming or transgender

improvement when it comes to how we treat menopause—and that the first step is a better understanding of ovarian biology.

For one, ovaries aren't just baskets of follicles. There are more than a dozen other types of cells gluing them together, all of which may influence how they grow and age. "And that's the big black box that nobody's looked at, really," she says. "I think we're really at a very primitive level of our full understanding."

―――

When Tilly's paper on ovarian stem cells came out in 2004, Woods was in graduate school studying the chicken ovary. She remembers being unsurprised to learn that ovaries would have regenerative powers. "It just seemed very logical," she says. "In my world, that's totally plausible." That's because those who study chickens have long known of this animal's peculiar ability: if you cut out its ovary, a new one can grow back, eggs and all. (Sometimes, it even grows a new testis.) But no one had thought about the ramifications for humans. "Now I kick myself because it was right in front of me," Woods says.

The ovary, it was clear, was dynamic. "Just look at the menstrual cycle," says Tilly, who holds thirteen patents related to fertility preservation and ovarian tissue bioengineering technologies (five of them shared with Woods). "But the key word is regenerative. Does it have inherent regenerative capacity? And the answer to that is one hundred percent unequivocally yes." In other words, stem cells were a hint that you should be able to harness that regenerative power to rebuild an ovary. As of the publication of this book, Tilly and Woods believe they are homing in on a second kind of ovarian stem cell—the kind that would give rise to new granulosa cells, not egg cells. With this last

youth who want to transition genders. They are supported by the Endocrine Society and the World Professional Association for Transgender Health, and have been deemed safe for children with precocious puberty by the USDA.

piece, they may soon be able to grow both granulosa and egg cells in vitro, and combine them to create an artificial ovary prototype.

In January 2021, in the lab she shares with Tilly at Northeastern University—the Laboratory for Aging and Infertility Research, informally known as the LAIR—Woods puts a thin slice of mouse ovary on the stage under the microscope. The computer screen shows a landscape of ghostly gray rivulets, like dry lakebeds on the surface of the moon. These are ovarian stem cells. Each one is giving rise to what looks like a translucent golf ball: a new egg cell trapped in time, caught forever in the act of being born.* For Woods, this is proof positive that the ovary is a dynamic, regenerative organ. "These are the cells that don't exist," she says. "But here they are."

She puts another petri dish under the microscope and points at a dark spot in the middle, with a fuzzy border—kind of like an Alka-Seltzer tablet dissolving in water. This is a spheroid, a group of cells working together that her lab has created in a culture dish. When she turns off the backlighting, the dark spheroid suddenly glows with four bright-green spots: ovarian stem cells, stained a fluorescent green, using the same jellyfish marker as Tilly used before. Each one is creating eggs and forming new follicles.

Most exciting to Woods is that the spheroid is both producing hormones and responding to hormone signals, meaning it has the potential to reattach the hormonal loop. This is proof of concept for the artificial ovary she's been working toward for a decade. It is also, she believes, an ability that exists within every ovary—mine, or yours.

Even setting artificial ovary technology aside, this work is challenging science to reimagine how the human ovary works—and rethink what the female body is capable of. Telfer, for instance, was initially "completely skeptical" of the new eggs idea. After Chinese scientists purified ovarian stem cells in 2009, however, she decided to take

---

* She's also made a time-lapse video of them dividing to the beat of the song "Moves Like Jagger" by Maroon 5.

another look. Her turning point came when Woods and Tilly offered to come to her lab in Scotland, ovarian stem cells in hand. They left her cells to culture, and she saw for herself that they were actively dividing and creating oocytes. "At that point I began to think, Yeah, there's something in this," she says. Eventually her lab isolated the cells themselves, and she went on to coauthor several papers with Woods and Tilly.

Now Telfer says researchers like Woods and Tilly are necessary for the field to advance. "The reality is, sometimes there's things there you don't see because you're not looking," says Telfer. "And this has opened our eyes that we need to start maybe taking a different tack." Though she still isn't sure whether these cells actually contribute to the oocyte pool in a normal ovary, she's keen to find out exactly what they may be doing.

It's the kind of openness that makes Tilly and Woods hopeful. At this point, there have been dozens of papers showing the existence of ovarian stem cells, while the papers seeking to disprove them are dwindling. The field, it seems, is finally waking up. "Slow and sure, people are coming around," Tilly says.

As women, we are not only made for reproduction.
—ELISE COURTOIS, ENDOMETRIOSIS RESEARCHER AT
THE JACKSON LABORATORY

out and the ruddy color faded from her cheeks, they reassured her she still looked great. Her husband, who was also her collaborator and co-director of her bioengineering lab at the Massachusetts Institute of Technology, provided a steadfast source of support.

Compared to Griffith's previous experience with reproductive disease, it was like night and day. Since puberty she had suffered silently from a common, painful condition called endometriosis. In it, cells similar to the lining of the uterus escape into the pelvis and take root, where they respond to the body's hormones, growing thicker before shedding and attempting to bleed. Breast cancer, on the other hand, was something that everyone recognized and empathized with. Compared to endometriosis, she likes to say, it was a walk in the park. "Not like a super beautiful day . . . like a stormy-day walk in the park," she adds. "But it was like, people understood."

The contrast wasn't only in how people treated her. It was in how doctors treated the two diseases. With breast cancer, doctors immediately biopsied her tumor, analyzed it, and categorized it so they could guide her into the right treatment. Tests checked for the presence of simple biomarkers—receptors for estrogen, progesterone, the protein HER2—that gave clues about how the tumor would progress and what treatment it would respond to. With endometriosis, there were no known biomarkers. There was no good classification system. And there were no treatment options other than surgery or hormone suppression, both of which had serious drawbacks. "There's no metrics," says Griffith.

Griffith was sick of hearing doctors describe endometriosis in terms of myth rather than science. She knew her disease could be explained in terms of data and biology. It just required viewing the uterus for what it was: not some mythical center of womanhood but an organ like any other, deeply connected to everything around it, trading in immune cells, stem cells, and vital fluids. In other words, it was part of a complex biological system. And as a bioengineer, that's how Griffith was trained to think: in terms of living systems made up of interlocking networks.

Endometriosis, like breast cancer, was not one disease but many, a hydra of many heads. To get a handle on it, you needed to look at entire networks of immune cells and how intervening in one pathway might affect all the others. She began talking to her husband, Dr. Doug Lauffenburger, who had been studying breast cancer for over a decade, about how to take a similar approach to classifying endometriosis patients.

Over the next year, Griffith held lab meetings from her hospital bed in between chemotherapy sessions, directing her lab to look for networks of molecular markers in endo patients. "We transformed our lab meetings, literally," says Dr. Nicole Doyle, a postdoctoral fellow in Griffith's lab at the time. "We just showed up for her chemo treatments and would sit there with her. That diagnosis had to adapt to her life, not the other way around."

Griffith's team started by analyzing peritoneal fluid, the fluid found inside the abdominal cavity where endometriosis lesions usually appear. They identified networks of inflammatory markers and used them to group together patients who tended to have more pain and worse fertility problems. In 2014, they published the first study to propose a way to categorize endo patients into subtypes—the first step in building a similar classification system that already existed for breast cancer. "That was really us together, because it was Doug's vision of systems biology but filtered through my practical connection to the clinic," Griffith says.

Throughout chemo, Griffith never seemed to waver in her positivity. When she shaved off her hair, she threw a lab party. Lauffenburger took it harder. For him, watching his wife suffer from this new foe, after battling the old one for so long, was torture. When it came to cancer, "I viewed it as a terrible thing," he says.

Griffith saw it differently. She took a curse and turned it into a gift. "It was a terrible thing, but a good thing, scientifically," she says.

⁓

Griffith started her career not in reproductive medicine but in tissue engineering, sculpting organs like liver and bone. She was an architect;

her materials were the building blocks of life. As one of the few women in her field, she made an effort not to draw attention to her gender. "I was working on all the things that guys were working on," she would say. "It didn't ever occur to me to work on a women's thing."

She grew up fearless, a tree-climbing Girl Scout in Valdosta, Georgia. From a young age, her parents instilled in her the idea that there were no limits to what she could achieve. She spent her time running barefoot outdoors, climbing trees, and earning her black belt in karate. When she was sixteen, she replaced the family's car radiator. In her family, "there was nothing we couldn't do, whether you're male, female, whatever," recalled her younger sister, Susan Berthelot. "We had a lot of confidence, and a lot of love, and a lot of freedom. A lot of freedom to take risks."

But when Griffith hit puberty, her body began imposing limitations on that freedom. Her period was a wrenching affair, bringing with it stomach-turning nausea, stabbing pain, and a heavy, unstoppable flow. When she was thirteen, her gynecologist prescribed her birth-control pills, a scandalous proposition—"in the South especially, it was not done," she says. Her mother, at a loss, gave her gin.

Unable to control what was happening inside her body, Griffith focused on what she could control: math, and building things. She went to Georgia Tech on a scholarship to study chemical engineering. But her physical problems only grew worse. She found herself failing tests when she was on her period, and going to the campus infirmary to get shots of Demerol, a powerful opioid. Once, during a chemistry class, the room started to spin. She tried to make it back to her dorm, but somewhere along the dirt path, she fainted. Another student, who she had a crush on, found her and drove her home. Too weak to open the door, she threw up all over his backseat.

By the time she went to UC Berkeley to get her PhD in chemical engineering, she had developed an elaborate period regimen: She wore all-black outfits, inserted three Super Plus tampons at once, and swallowed more than thirty Advil tablets a day. Yet most doctors were less interested in her symptoms and more interested in how she managed to take so many painkillers without getting a stomachache. When she consulted one male doctor, he took a look at her pixie cut and wiry, athletic build and diagnosed her as "rejecting her femininity."

"I felt like I was being gaslighted," she says.

Her real diagnosis came by accident. It was November 1988, and she had recently moved to Cambridge, Massachusetts, to complete a postdoc in a tissue engineering lab at MIT and live with her first husband. After prodding her doctor about her pain for six months, an ultrasound revealed a small cyst on her left ovary. Draining it would be a same-day procedure, in-and-out, her doctor said. But when she woke in the Brigham and Women's Hospital in Boston, it was the next day. She looked down and saw a row of staples along her midriff, holding together a six-inch incision.

While she struggled to gain clarity, her gynecologist came into the room to explain the situation. Her husband was already sitting beside her.

The problem was not Griffith's ovary, her doctor said, but her uterus. She had a chronic disease called endometriosis. It was common, striking as many as 1 in 10 women, trans men, and nonbinary people who menstruate. Because nobody had taken her pain seriously enough to examine her, Griffith's disease was particularly advanced: her ovaries, bladder, and intestines were all fused together with a sticky, speckled tissue that resembled the lining of her uterus. They'd cut out as much as possible, and burned off the rest. There was little else they could do.

Griffith could only hold on to one thought: she had a real disease. "To have someone tell me something's wrong with me, it was a huge relief," she says.

Her doctor laid out two options: She could go on Danazol, a powerful steroid that would stop her body from producing estrogen and put her into a state of temporary menopause. Or she could get pregnant.

She recalls her then–husband answering for her: "We'll have a baby."

Griffith, though she had always wanted children, opted for the Danazol. Two years later, she left the husband and began her career in tissue engineering, a brand-new field that wielded the power to sculpt new organs from living cells. She developed an artificial liver, figuring out how to create polymer scaffolds in the lab and seed them with living blood vessels. In 1997 she created an iconic creature called the "earmouse" by injecting a human ear-shaped scaffold with cartilage from a cow's knee and growing it on the back of a lab mouse. Her work in the lab of MIT chemical engineer Robert Langer helped jumpstart tissue engineering, which today provides artificial skin and organs to millions of burn and injury victims. "She published, I'd say, some of the seminal papers in this field," says Langer.

Yet she never thought of turning her organ-making skills to the uterus. "Psychologically, it wasn't something I wanted to think about," she says. "I just wanted to pretend like it wasn't happening."

Endometriosis was her burden, work was her escape, and never the twain shall meet.

———

In the 1980s, when Griffith was first diagnosed, medical textbooks had dubbed endometriosis "the career woman's disease." Doctors stereotyped their patients as "underweight, overanxious, intelligent, perfectionist, white, of high social and economic standing, between 30 and 40 years, with regular menstruation and ovulation, who regularly delay childbirth," according to one study that examined these biases. They commonly prescribed marriage and pregnancy as "cure," with the medical reasoning being that, since hormones can trigger the lesions to grow, tamping down the female hormone cycle could relieve women of the disease. Even though this logic has been roundly refuted, doctors still recommend pregnancy as treatment today.

Given that endometriosis was seen mainly as disease that robbed women of their fertility, pregnancy "was almost viewed as a two-for-one benefit," says Dr. Elizabeth Stewart, who performed Griffith's first surgery. "It's clear there was some sexism in the approach to endometriosis then. I think there's still some now."

Yet the idea that Griffith and other women with endometriosis were somehow to blame for their own suffering goes back further than that. To understand why, we have to return to ancient Greece.

To Greek physicians, the uterus was no ordinary organ but a beast that prowled, hungry for sex and motherhood. "An animal within an animal," wrote the second-century physician Aretaeus of Cappadocia. "In a word, it is altogether erratic." (The penis, too, was referred to as an animal, so this concept wasn't all that uncommon in Greek times.) Compared to man, woman was wetter and spongier of flesh, yet her uterus was light and dry. This meant it was always on the hunt for moisture, a quest that brought it in contact with other internal organs. When it didn't achieve its aims, it grew sullen and melancholy, causing mayhem throughout a woman's body. It squished up against the intestines, lungs, and heart, which could make her faint, spasm, or choke.

This "extremest anguish," Plato wrote, was caused by a woman allowing her womb to remain barren too long after puberty. The uterus, Hippocratic texts declared, was "the origin of all diseases."

The most commonly described condition was *hysterikê pnix*, meaning "suffocation of the womb," when the uterus lurched up and down the body. Widows and young, unmarried women were especially vulnerable to this malady. But when the womb wandered off, it could be tempted back by scent. To attract it upward, a physician waved sweet-smelling substances in front of the lady's nose; to bring it downward, he instead put them near her nether regions. Other treatments were less endearing. One was fumigation, in which hot air was blown up through a reed into a woman's vagina.* Another was bandaging the abdomen tightly to keep the womb in place. Yet another was bloodletting, achieved by applying leeches to the cervix or labia.

Today, the word "hysterical" is often used a way to write off women as irrational and overemotional. But in ancient Greece, it was a medical diagnosis. The ultimate cure was always the same: the holy trinity of marriage, sex, and pregnancy. Intercourse, it was thought, introduced moisture and stirred up the body's fluids. Babies, meanwhile, were the uterus's raison d'être: they weighted it down and kept it in its proper place. The Greeks described the uterus as an oven in which the seed of the male is cooked to form new life. But an oven must be occupied, or it overheats. Similarly, if a woman stayed barren too long, she would be susceptible to womb movement and its accompanying sickness. A womb, like a woman, must be occupied.

You'd have thought the advent of human dissection would have cleared up the notion that the womb wanders. Not so. In the second century, Galen, while studiously ignoring the clitoris, confirmed that the uterus was not actually a free-roaming organ: it was anchored to the pelvic walls by flexible ligaments, or membranes. Galen concluded that uterine diseases were actually caused by these ligaments swelling with blood, male seed, or unfertilized female seed, which decayed in the womb and produced harmful vapors. Thanks to a woman's wetter nature, she needed to bleed every month to get rid of her body's excess fluid and avoid such a fate.

Those who came after Galen knew about the uterine ligaments, yet many simply incorporated this new anatomy into the old framework. Some said the womb still moved, but was pulled back by extra-stretchy ligaments. Others continued to recommend scent therapy, with the rationale that it could relax or constrict the ligaments. The idea of the wandering womb traveled from West to East, and held sway over medicine for centuries, writes Helen King, a

---

* Not unlike today's steam douches.

professor of classical studies at the Open University in the UK, in her book chapter "Once Upon a Text: Hysteria from Hippocrates." Even Victorian-era smelling salts borrowed from that same logic, promising to revive a swooning woman by coaxing her uterus back to where it belonged.*

Why did the idea of the wandering womb remain such a resilient concept, even after anatomical progress had proved it wrong? King, who studies attitudes toward menstruation in ancient Greece, has a theory: "It's a very useful way of keeping women in their place," she says. "It keeps women focused on childbearing. It means other options become a threat to their health."

While the explanations for hysteria shifted, one thing remained constant: It was a biological disease, with biological causes. Although it tied a woman to her reproductive biology, it at least gave her a solid diagnosis, a name for her pain. By the twentieth century, that would start to change. Soon, medicine began thinking of hysterical women not as patients afflicted with a bodily illness but as neurotic women whose problems were all in their heads. By 1900, the word "hysteria" had lost virtually any connection to the uterus.

What accounts for this dramatic shift? For that we can, once again, thank Freud.

⌐

A Frenchwoman leans backward in a swoon. Her eyes are closed, her corseted breasts thrust forward. A crowd of bearded gentlemen lean forward in their seats. In the center of the scene, depicted in the 1887 painting *Une leçon clinique à la Salpêtrière* (A clinical lesson at the Salpêtrière), a gray-haired man in a black suit gestures toward her. He

* The sixteenth-century physician William Harvey, though he made great strides in understanding how blood circulated in the body, also wrote that hysteria was brought on by "unhealthy menstrual discharge" related to "being too long unwedded."

is illustrating her stance: passive and objectified, draped over the arm of an assistant, in the *arc en circle*, the classical posture of the hysteric. This, for decades to come, would be the iconic image of hysteria.

The gray-haired man in the painting is Jean-Martin Charcot, a neurologist and director of the Salpêtrière, a mental asylum and teaching hospital outside Paris. Charcot would gain renown for identifying diseases like multiple sclerosis, aphasia, Tourette's syndrome, and ALS, which in France is still sometimes known as Charcot's disease. But his pet interest was always hysteria. In the 17th century, hysteria had nearly died an undignified death, becoming wrapped up not in science but in witchcraft, demons, and sorcery. Charcot rescued it from the ashes. While others ridiculed it as an affliction of witches and malingerers, he argued that hysteria was in fact an organic disease—just one that lived in the brain, not in the reproductive organs.

By Charcot's time, psychiatric hospitals like the Salpêtrière had spread across Europe, many of them filled with so-called hysterics. The standard nineteenth-century treatments were every bit as brutal as those of ancient Greece: leeches, pills, arsenic, opiates, induced vomiting. Charcot had his own methods. Every Tuesday, within a five-hundred-seat amphitheater he had created for this purpose, he would demonstrate a hysterical attack by hypnotizing a patient using the sound of a gong or tuning forks. He had come to think of these attacks as a finely choreographed dance, taking fifteen to twenty minutes, in which the victim went through the same steps: a rigid posture, standing straight up; grand, circuslike gestures of the limbs (Charcot was a great fan of the circus); and finally, fainting backward in a dramatic arch. He illustrated these steps with colored chalk on the blackboard.

Charcot's demonstrations were dramatic and vaguely erotic, filled with writhing and moaning. He claimed to be able to stop an attack by using experimental methods like hypnotism, "animal magnetism," and electricity. (He also believed hysterical episodes could be triggered or stopped by pressing on the ovaries, and invented a brutal-looking device called an "ovary compressor" to do just that.) Ultimately, his performances would be revealed as fraudulent, and hysteria would disappear as a diagnosis in Paris. But that didn't stop one young neurologist from taking up its mantle.

In 1885, one of Charcot's audience members was a medical student named Sigmund Freud. Freud, who had been working in the neurology lab of Ernst Brücke comparing the brains of frogs, crayfish, and lampreys, had come to Paris for six months to study under Charcot. Like others in the audience, he was stunned by what he saw. He was especially intrigued by Charcot's work on male hysterics, and his attempts to show that the disease stemmed not from the uterus but from some invisible injury to the nervous system. He would take these observations one step further: in his opinion, the heart of this disease was not a physical injury but a "psychological scar produced through trauma or repression" that manifested in physical symptoms.

Freud returned to Vienna eager to convince his colleagues of the merits of hypnosis in treating hysteria. Yet when he began lecturing on male hysteria, he was met with ridicule. "But, my dear sir, how can you talk such nonsense?" he was told by one incredulous older surgeon. "Hysteron (sic) means the uterus. So how can a man be hysterical?" Freud disagreed. It was a mistake, he wrote, to link hysteria to the womb. The word itself was a "precipitate of the prejudice, overcome only in our own days, which links neuroses with diseases of the female sexual apparatus."

Hysterical symptoms like nervous coughs, painful breathing, migraines, anxiety, and muteness could affect men or women, he argued. "Hysteria behaves as though anatomy did not exist or as though it had no knowledge of it," he wrote in 1893. Freud had reversed the symptoms: rather than menstrual problems causing anxiety and neuroses, now it was neuroses and anxiety manifesting as biological symptoms. The wandering uterus was no longer literal, but metaphorical.

Hysteria, for Freud, was a stepping-stone. Once he had wrested the disease away from medical doctors, he was able to get on with his true project: showing that all neuroses had their basis in the mind, and specifically, in traumatic sexual memories or sexual conflict. Hyste-

ria served as a proof-of-concept for his argument that, by making his
patients confront traumatic memories, he could rid them of trouble-
some physical symptoms. In 1895, he and a colleague, Viennese physician Dr. Josef Breuer, published *Studies on Hysteria*, where he first laid
out his sexual thesis. "Hysterics," they concluded, "suffer mainly from
reminiscences." In other words, it was all in their heads.

It is perhaps no surprise that the turn away from biological causes
and toward blaming women for their illnesses coincided with the rise
of first-wave feminism in Europe and the fight for suffrage. As women
became more visibly engaged in the outside world, leaving the traditional sphere of the home, doctors began fretting that this unnatu-
ral assertiveness was leading to their ill health. Higher education and
careers, they feared, might siphon blood from their uteruses to their
brains. But their pronouncements soon had an air of blame to them.
"Cures" like hysterectomy, ovariotomy, and pregnancy now began to
sound more like punishment.*

Instead of blaming women's uteruses, Freud cut to the chase and
blamed women.

———

Freud was never particularly interested in the uterus itself, except as the
tabula rasa on which he could build his psychic empire. Besides a few

---

* Some did recognize that these so-called solutions weren't cutting it. One was Lydia
E. Pinkham, a housewife-turned-entrepreneur from Lynn, Massachusetts, who saw
that male doctors failed to understand—or lacked empathy for—the needs of their
women patients. "What does a man know about the thousand and one aches and
pains peculiar to a woman?" she wrote in her widely distributed pamphlet, "Treatise
on the Diseases of Women," first published in 1901. In frank, easily understood language, it laid out reproductive anatomy and the biology of ovulation, fertilization,
and pregnancy, woman to woman. Unfortunately, the main point was to sell Lydia
Pinkham's Vegetable Compound, a proprietary mix of concentrated herbs and roots
that claimed to ease the pains of menstruation, menopause, and innumerable other
uterine ailments. It was later revealed that the tonic's main ingredient was alcohol.

instances of men wishing to give birth, and one woman who experienced a "hysterical pregnancy" in which she gave birth to nothing, the uterus scarcely shows up in his texts. Yet while gynecological anatomy hardly influenced his theories, his theories would deeply shape gynecological medicine.

Like Charcot, Freud considered the disease as an equal-opportunity neurosis, one that afflicted men as frequently as women. He even once referred to working through his own "little hysteria." Yet the vast majority of his hysteria patients—and almost all of the ones who formed his case studies—were women. (Men with identical symptoms generally got a diagnosis like neurasthenia or shell shock, today known as PTSD.) Women, Freud believed, were by nature more prone to nervous disorders, because of the sexual conflict they faced along their tortuous path to womanhood. And it was women who would bear the brunt of hysteria's legacy.

In 1980, hysteria was finally deleted from the Diagnostic and Statistical Manual of Mental Disorders. Yet it lived on in a group of diagnoses known as the "psychosomatic" disorders. "Hysteria dressed up in modern garb," as journalist Maya Dusenbery called them in her 2017 book *Doing Harm,* these diagnoses were all considered "female" ailments: they were diagnosed ten times more often in women as in men. In reality, Dusenbery argues, women suffer disproportionately from little-known conditions like chronic fatigue syndrome, possibly due in part to differences in their immune systems or other biological differences. Yet when doctors can't promptly explain their symptoms, they get lumped by default into one of these psychological categories.

Meanwhile, diseases that really do stem from the uterus—like endometriosis—still get dismissed as Freudian problems of the psyche.*
When Abby Norman first went to her doctors with symptoms of endo-

---

* Some scholars argue that hysteria, far from being a made-up disease, has always actually been endometriosis in disguise. "If so, then this would constitute one of the most colossal mass misdiagnoses in human history, one that over the centuries has subjected women to murder, madhouses, and lives of unremitting physical, social, and psychological pain," write the Nezhat brothers, three endometriosis surgeons

metriosis as a college student, they dismissed her theories. "You were probably molested as a child and this is just your body's way of trying to handle it," one told her. "This is all in your head," said another. Once she was finally diagnosed in her twenties, doctors assumed that her priority was having children.

"The things that actually did concern me—the pain, the nausea, the complete loss of everything that I loved and that made me happy (food, dance, sex)—didn't seem to carry the kind of weight that concerns about my fertility did," she wrote in her 2018 memoir, *Ask Me About My Uterus*. "How, I wonder, did the doctors expect me to get pregnant if I couldn't have sex? What if I had said, 'Okay, fine, I'll have a baby—but how, pray tell, shall I go about it when sex is excruciatingly painful and I can't tolerate penetration long enough to be fertilized?' Why wasn't it enough that I was a young woman who wanted to be sexually active, but couldn't be?"

The presumption that her end goal was to be a mother was so deeply entrenched that at times, her doctors didn't even bother to ask. In her first exploratory surgery for endometriosis, a surgeon found a large cyst that had displaced her ovary and twisted the adjoining Fallopian tube. Rather than remove it, the surgeon only drained it, so as not to threaten her fertility. The pain came back within weeks. Norman was not overly concerned about her fertility; she just wanted to be free of pain. Her disease had wrecked relationships, stopped her from going to college, and made her feel constantly ashamed. Yet medical professionals made her feel like she'd brought it upon herself—for wanting a career, for wanting sex, for not wanting children. For not, as Freud would have said, adapting to her role as a woman.

There's a reason some scholars have deemed endometriosis "the new hysteria."

---

from Iran, in a 2012 paper. "The number of lives that may have been affected by such centuries-long misdiagnoses is staggering to consider."

From the outside, Dr. Linda Griffith was unstoppable. A fast-talking dynamo who showed up to campus riding a Kawasaki motorcycle and sporting a leather jacket, she cut a memorable figure in the minds of her colleagues.* "Bountiful amounts of energy, totally brilliant, preternaturally young-looking, and just a phenomenon," remembers Harvard geneticist Pardis Sabeti. "Like a thunderstorm. She's the most highly energized person I think I ever met," says MIT toxicologist Steven Tannenbaum. "She's just supercharged."

None of her colleagues knew, however, what was going on under that energetic exterior. Throughout the '90s, Griffith underwent one invasive surgery after another. Yet her rogue uterine tissue kept growing back. It surrounded her bowels and ureters, squeezing them down. By the fifth surgery she could no longer tolerate crossing the Charles River. A knot would form in her stomach as she remembered all the painful hours spent in the hospital. At the same time, she was taking strong drugs, like Lupron, a hormone blocker that gave her short-term memory loss. While teaching a class on thermodynamics, she would be forget terms like "heat transfer."

She still hadn't given up on her dream of having children. In 1994, she helped recruit Lauffenburger, a systems biologist at the University of Wisconsin, to work alongside her as the head of MIT's new biological engineering department. After working together in the lab, the pair fell in love and quietly got married.† In 1997, they went through several rounds of IVF in the hopes of conceiving a child, but none of the embryos took, probably because her disease was already so advanced. Today, three carved stone cherubs hang above the doorway to her

---

* Today, when she isn't running the six miles to the Cambridge reservoir and back, she can be found bouncing on her hydraulic pogo stick, which she taught herself how to use during the pandemic.

† Some of their lab students are still taken aback to find out they are a couple.

kitchen: a gift from Griffith's mother, to commemorate the embryos that never were.

In September 2001, the day after her fortieth birthday, Griffith woke up with an attack of stabbing abdominal pain. Her doctor gave her an opioid, but it barely masked the agony; she had to mix it with two glasses of wine. The following morning was September 11. While the nation looked on in horror as the Twin Towers fell, Griffith rushed to the hospital in a fog of painkillers and underwent a hysterectomy with her surgeon, Dr. Keith Isaacson. The choice had been made for her: She had to get rid of her uterus, the center of her pain but also her hope for children. "There was no decision. It was hysterectomy or death," she says.

Finally, she thought, she could close the chapter on her endometriosis and move on with her life. But her uterus wasn't done with her yet. In 2005, the disease came back, requiring two more surgeries.[*] Afterward, she tried her best to avoid the thought of motherhood. She would come up with excuses not to attend a work dinner where a colleague's wife would be bringing her newborn. She knew that to stay on top of her career and keep herself sane, she had to keep the darkness out. "The hellmouth will open at the dinner," she says, "and you've got to keep the hellmouth closed."

⌣

The turning point came in 2007, when a member of the MIT Corporation's board of trustees, Susan Whitehead, asked Griffith to speak at a Women in Science and Engineering luncheon about how her work on tissue engineering could benefit women. At first, Griffith was annoyed. "I was super not into the women thing," she says. "I

---

[*] Given that endometriosis is, by definition, a disease that occurs outside the uterus, removing the uterus rarely solves the problem for good. Often, more lesions are hiding in other parts of the pelvis, sometimes burrowed deeply into tissues. More than half of women who undergo hysterectomies for endometriosis have recurrent pain, and many have to undergo further procedures.

just tried to stay out of it because I wasn't part of the narrative." But Whitehead was a friend, so she agreed.

Toward the end of the event, the moderator asked Griffith where she saw herself and her work in ten years. She found herself thinking about her niece, Caitlin, who had just been diagnosed with endometriosis after years of being told that her symptoms stemmed from stress.

She found herself blurting it out: "I have a chronic disease called endometriosis," she began. "My niece who's sixteen was just diagnosed. And there's no better treatment for her—thirty years younger than me—than there was for me when I was sixteen." She herself had just had her eighth surgery for the disease. But it was her niece who "made lava shoot out of my head," she says.

When it came to making liver and bone, "so many other people could do them. But there was this one thing only I could do." She had recently been awarded a prestigious MacArthur "genius" grant, which came with half a million dollars for any research project. Now she knew what she was going to do with it. In 2009, she used it toward opening the Center for Gynepathology Research at MIT, the only engineering lab in the nation to focus on endometriosis and a related yet even less known condition, adenomyosis, in which similar tissues grow within the muscular walls of the uterus.

During the launch event for the center, Padma Lakshmi, host of *Top Chef* and co-founder of the Endometriosis Foundation of America, lamented the lack of research on such a devastating disease. "I have to say, I'm really shocked that it's the first research center of its kind in America," she said. "That is stunningly bad news on the one hand, that she's the first one doing it. On the other hand, better late than never. Thank God for Dr. Linda Griffith."

⎯

Most labs devoted to women's diseases are accompanied by obvious symbols of womanhood: a rose, a tulip, an hourglass silhouette. Not

Griffith's. Tucked away in the building for biological engineering, the Center for Gynepathology Research is marked only by the letters CGR in red and black, the *G* formed from a curved arrow representing the hand of the engineer. "We needed something that wasn't all pink and flowers," Griffith says, in her slight Georgia drawl. "We really thought it should be, like, 'This is science.'"

Like Griffith herself, her lab speaks the genderless (some would say masculine) language of science and engineering. It's part of her push to change the conversation around endometriosis from one of women's pain to one of biomarkers, genetics, and molecular networks. "I don't want to make endometriosis a women's issue," she told the *MIT Technology Review* in 2014. "I want to make it an MIT issue."

In her lab, she has begun growing uterine organoids—tiny domed droplets, with glands that look like swirling craters—from the uterine cells of endometriosis patients. Placed in a gel made to mimic the uterine environment and fed the right nutrients, these cells spontaneously form structures that resemble the human uterine lining, growing and shedding in response to hormones. These "patient avatars" are ideal tools for testing potential new treatments for the disease: Biologically, they are closer to human uterine cells than those of mice, as mice don't naturally menstruate. And they enable researchers to sidestep some of the ethical issues that arise with human trials.

Her research highlights what a remarkable organ the uterus truly is—and not just during its signature function, pregnancy. Humans, unlike almost every other mammal, grow their entire endometrium—the womb's inner lining—once a month, whether or not a fertilized egg takes hold. If no egg appears, they shed it.

Picture the womb as a small orange, and the pith would be its lining—a plush, living bedding for a potential embryo. Each month or so, triggered by a drop in the hormone progesterone, this lining sloughs itself off and builds itself anew. Immune cells rush to the site to heal the wound. Connective cells that line the uterus differentiate into new lining, complete with delicate, spiraling blood vessels. The process repeats itself, swiftly, scarlessly, without a trace of injury, again and again, up to five hundred times in a woman's life. "How the body can coordinate that is extraordinary," says Dr. Hilary Critchley, a reproductive biologist at the University of Edinburgh.

To capture these systemic interactions, her team is seeding their models with blood vessels, nerve cells, and immune cells. They hope to eventually connect them to models of the liver, bone, and gut. Clearly, Griffith sees the uterus far differently than the Greeks did: not as the center of female frailty but as a powerhouse of renewal and regeneration. Dynamic, resilient, and prone to reinvention, this organ offers a window into some of biology's greatest secrets: tissue regeneration, scarless wound healing, and immune function. "The endometrium is inherently regenerative," she says. "So studying it, you're studying a regenerative process—and how it goes wrong, in cases."

———

As it turns out, Griffith's in vitro models are sorely needed. When it comes to understanding endometriosis—and menstruation in general—science lacks good animal models.

The art of growing and then sloughing off the entire uterine lining is a rare quality throughout the animal kingdom, limited to a handful of primates, four species of bats, and a couple shrews. Even fewer experience menstrual disorders like endometriosis. Altogether just 84 species—1.6 percent of all placental mammals—are known to menstruate. If you look at a phylogenetic tree, they're all over the place, suggesting that uterine bleeding has evolved at least three times independently. This underscores a fundamental mystery: Menstruation, with all its requirements and regularity, is why the human uterus is so dynamic. But it is also costly, requiring an animal to shed an entire organ and regrow it every month. So why would any animal bleed?*

---

* Humans may not be quite as exceptional as we think. The Cairo spiny mouse, a

One of the most common explanations is the idea that the female body needs to get rid of something dirty or harmful. In the 1920s, a doctor named Béla Schick theorized that women had special toxins in their menstrual blood, which he called "menotoxins." His questionable experiments suggested that menstruating women sweated toxins from their skin that made flowers wilt and die. Although none of his findings were replicable, others latched onto the idea, arguing that menstruating women did indeed wither plants and spoil beer, wine, and pickles. Even today, many theories about menstruation come from the idea of the vagina as dirty and in need of cleansing: In 1993, a physician and mathematician named Margie Profet made waves when she suggested that menstruation's function is to "defend against pathogens transported to the uterus by sperm."[*]

The real explanation may be not about getting rid of something harmful but defending against harm in the first place, says Dr. Günter P. Wagner, a Yale researcher in the ecology and evolutionary biology department who studies the evolution of menstruation.

Consider that motherhood isn't all warm and fuzzy. It's a fight for resources, often waged brutally between mother and offspring. Since an offspring holds only half the genome of its mother, its evolutionary interests aren't exactly the same as hers—and sometimes, they directly conflict. Evolutionarily speaking, the goal of the fetus is to suck as many resources as possible from Mom, whom it basically sees as its personal Giving Tree. Mom's goal, by contrast, is to survive her pregnancy and limit the aggressiveness of her offspring. The evolutionary

---

rodent native to the Middle East, doesn't look like much—just "a little bottlebrush that has eyes and a tail," says Nadia Bellofiore, a researcher at Monash University who works with a colony of them. But it is the only rodent that we know of that bleeds. Bellofiore has found that, like us, this "humanesque rodent" ovulates spontaneously, sheds its lining, and marshals immune cells to repair the wound. Interestingly, spiny mice are also known for their regenerative powers: they regenerate skin and hair follicles.

[*] Though you could say that this at least shifted the blame to the penis.

tug-of-war that takes place between mother and child's genomes is called "maternal-fetal conflict."

Animals that menstruate have particularly conflict-riddled relationships with their offspring, Wagner says. They tend to have more invasive fetuses and placentas, which burrow deeply into the mother's body to gain access to her nutrients and blood supply. This poses an existential threat: In blurring the boundaries between the mother's body and her offspring's, the fetus runs the risk of siphoning off too many resources and weakening or even killing its host.

Fortunately, Mom has some tricks up her sleeve. The key event in menstruation is not bleeding, but the differentiation of the uterine lining. Over a period of about three days, uterine cells called fibroblasts transform into what are known as "decidual cells"—meaning that they eventually fall off, like the leaves of deciduous trees. While these cells are necessary for an embryo to implant, they simultaneously create a matrix that is more difficult to penetrate. They also help tamp down the inflammatory response that occurs when a fetus implants, an event akin to a wound. All of these developments make sure the embryo burrows deeply enough to stay viable, but not so deep that it harms the mother.

In most species, this crucial differentiation happens only when an embryo appears. But in menstruators, it happens about once every month, spontaneously. (Menstruating animals ovulate, or release eggs, spontaneously—as compared to animals who ovulate in response to light and temperature, like frogs, or copulation, like dogs.) "You don't want to be defenseless when this pesky embryo is coming along," says Wagner. You want to be ready for it—"sort of like a standing army." These animals get a head start by erecting their defenses with every ovulatory cycle—no fetus required.

Once that lining has differentiated, and once the body realizes there is no embryo, it has nowhere to go but down. The progesterone drop causes the blood vessels to violently die, killing the surrounding tissue and causing the rest to disintegrate and exit the body through the vagina.

So what really links animals that menstruate? Evolutionarily, they're Type-A planners. They anticipate conflict, priming the uterus and shielding themselves should an unwanted visitor happen to implant, rather than waiting until it's too late. Mom's body does this independently, regardless of whether a male, or fetus, comes along. There may be another advantage to regular menstruation: The uterine lining could play a role in sensing the "quality" of the fetus and deciding whether a prospective embryo should live or die. By taking into account chromosomal errors, aging sperm and eggs, and other quality-control issues, the mother might summarily eject an embryo that isn't worth the investment. The uterus may even learn from its mistakes and adapt to new conditions.

Many researchers argue that the remarkable dynamism of the uterine lining is a double-edged sword. In the past, Critchley points out, women menstruated only around forty times in their lifetime, and spent the rest of the time pregnant or nursing. Today, the average Western woman menstruates up to five hundred times—meaning it's statistically more probable that some step in the intricate process will go askew. Consider endometriosis: Taken out of its natural context of the womb, the dynamic nature of the uterine lining proves catastrophic, intent on executing its life cycle in places where it caused scarring, pain, and inflammation.

Others counter that this logic is just the modernized version of Hippocrates's wandering womb: it presumes that women's uteruses are set up for disease, and that pregnancy is protective. There's no reason that the frequency of menstrual cycling should be inherently pathological, says Dr. Kate Clancy, a biological anthropologist who studies reproduction at the University of Illinois Urbana-Champaign. There are other changes to modern women's bodies that deserve deeper investigation, including external factors like toxins from the environment that have been linked to endometriosis. Perhaps the problem is not in women's heads, or even their pelvises, but in the world they inhabit.

"I'm increasingly thinking that it's not a system flaw," Clancy says. "At a certain point we need to start to think about this with the same rigor we would if this was a cisgender male body that we were exploring."

That starts with understanding the basic mechanisms of menstruation. Shedding light on processes like uterine differentiation will help reveal what makes the cells of endometriosis different from other uterine cells and, ultimately, help scientists disrupt the process. It isn't a woman's lot to suffer just because she isn't pregnant. We just haven't been asking the right questions about how the uterus truly works.

Nor, it seems, have we been asking the right questions about endometriosis.

———

One of the first doctors to systematically probe the origins of endometriosis was Dr. John A. Sampson, a twentieth-century gynecologist who practiced in Albany, New York.* Sampson had grown fascinated by a common yet mysterious condition in about 1 in 10 of his female patients. When he opened up their pelvises, he found "chocolate cysts" attached to their ovaries and uteruses—named for their contents, which resembled chocolate syrup. To figure out what was happening, he began intentionally scheduling hysterectomy surgeries while women were on their periods. He removed their uteruses, injected the arteries and veins with red and blue dyes, and inspected them under the microscope. That's when he noticed something strange.

Most menstrual blood flows downward, through the vagina. But Sampson observed that some also escaped upward through the fimbriae-tipped ends of the Fallopian tubes, into the fluid-filled pelvic cavity. Like cuttings from a tree, he mused, this escaping tissue

---

* He has an artery named after him, Sampson's artery, otherwise known as the artery of the round ligament of the uterus.

could seed the pelvis and spread to other organs. This, he believed, was the origin of the chocolate cysts. He called them "menstruating organs," which grew and then attempted to shed in response to hormones. Some were small and superficial, but others invaded deep into the pelvic tissue, or even into the uterine wall, similar to a cancer. The result was pain, irritation, more bleeding, and, most distressingly to doctors, infertility.

Sampson's explanation for how endometriosis takes root relied on the idea of a literal wandering womb, a reproductive system gone haywire. He considered endometriosis a disorder primarily of the uterus and ovaries—a disease of menstruation, reproduction, and women. He noted that endometrial cysts "affect women in a most valuable period of their lives, usually from 30 years of age to the menopause," and was one of the first to assert that pregnancy seemed to have a beneficial effect.*

Yet this approach never yielded a definitive answer. In 1940, Sampson concluded that the disease remained "tantalizingly alluring and elusive."

It still is, today. Despite the fact that this disease has been known for more than 150 years and afflicts at least 200 million women and other menstruating people worldwide, nearly every scientific paper on endometriosis starts out with the words "mysterious," "enigmatic," or elusive." "There are few diseases in gynecology that are as enigmatic as endometriosis," begins one 2010 article in a gynecology journal. "It remains a riddle that baffles researchers and clinicians." Surgeons sometimes refer to endometriosis as "the pelvic chameleon" or "the

---

* Later OB/GYNs put it more baldly: "Nature (since the beginning of time) has employed an efficient prophylactic and curative measure for endometriosis, i.e., pregnancy," wrote Dr. Clayton T. Beecham in 1949. "It is noteworthy that the frequency with which the diagnosis of endometriosis is made parallels the increased use of contraception, the emancipation or rise of womankind to careers and/or late marriage with late childbearing."

great pretender"—phrases that invoke the dissembling language of hysteria itself.

But is the disease really so elusive—or have we just been looking at it the wrong way?

A narrow focus on fertility may be one reason so little has changed when it comes to managing endometriosis. Getting a diagnosis usually requires a surgical procedure to confirm the presence of lesions, meaning that women wait years, if not decades, for an official diagnosis. Once they have a diagnosis, the two main options are still surgery to cut or burn away the lesions—which often grow back—or shutting down the reproductive system by depriving it of hormones. Newer drugs like Orilissa, a partial estrogen suppressant released in 2018, still rely on these same mechanisms, says Dr. Linda Giudice, a reproductive endocrinologist at UC San Francisco who studies the biology of the uterine lining. "It's not something super novel," she says. "It's a variation on a theme."

Endometriosis patients—like the nineteenth-century women who had their ovaries cut out—were once considered victims of their own delicate, flawed reproductive systems. Increasingly, researchers like Griffith are seeing it differently. To them, endometriosis is far larger than a "women's disease": it is a systemic disease of inflammation, affecting nearly every organ system. "We need to address this as a disease that affects many aspects of the body, whether it's inflammation, a dysfunctional immune system, inflammatory bowel syndrome," says Elise Courtois, a researcher who studies the genetics of endometriosis at the Jackson Laboratory. "As women, we are not only made for reproduction."

⸺

One enduring mystery of endometriosis is how lesions can appear in places as far-flung as the lungs, eyes, spine, and even the brain. The answer may have to do less with the uterus and more with regenerative processes happening throughout the entire body, notes Dr. Hugh Taylor, chair of the department of obstetrics, gynecology, and reproductive sciences at Yale School of Medicine. Taylor investigates whether stem cells, which are plentiful in the uterine lining, could contribute to the spread of the disease by circulating throughout the body. Stem cells outside the uterus may also play a role: in women with uterine damage, stem cells from bone marrow flow in to repair the damage.*

Chronic, low-level inflammation of the uterus may also contribute to the origins of the disease. Dr. Peter Gregersen, a rheumatologist, and Dr. Christine Metz, an immunologist, both at Northwell Health's Feinstein Institutes, have spent five years developing a simple diagnostic test for endometriosis using menstrual blood. When looking for a biomarker to base their test around, the pair put regular uterine cells into an inflamed environment and found that they transformed. They became stickier, more invasive, and worse at decidualizing—exactly like the uterine cells in women with endometriosis. These invasive cells, Gregersen noted, were similar to the ones he had previously studied in inflammatory diseases like rheumatoid arthritis and lupus.

If Gregersen and Metz's hunch is right, it could mean that anti-inflammatory drugs currently used for rheumatoid arthritis might be repurposed to prevent endo from taking hold in some women in the first place. "The endo community is very dominated by people who think this is an abnormality of hormonal regulation," said Gregersen. "And I mean, it may well be. But I don't think that's the whole story."

As for where all that inflammation might come from: Dr. Kevin Osteen, an obstetrics and gynecology researcher at Vanderbilt who has worked with Griffith, studies how early exposure to environmental toxins might lead to uterine inflammation and, ultimately,

---

* Because uterine stem cells are relatively accessible, they could also be a boon to regenerative medicine. Taylor has shown that, like other stem cells, uterine stem cells can be grown into new neurons and insulin-making cells to treat diseases like Parkinson's and diabetes.

endometriosis. Osteen began focusing on the disease in the 1980s, when he was leading Vanderbilt's fledgling IVF program and realized that many of his patients suffered from it. Since then, he has come to believe that the key to halting endo is tackling the inflammation associated with its early stages, long before it progresses to infertility. "In my mind, understanding the immunological origins of endometriosis opens the window to prevent the disease from even developing," he says.

Osteen has also found that the pollutants he studies, dioxins, lead to similar inflammatory pathways in both men and women. Though they don't lead to endometriosis in men, they can cause other problems with fertility and testicle function, and can be passed down to daughters. "It's not at all just a woman's problem," he says. "It needs to be looked at much more broadly than that."

Meanwhile, Griffith brings something else to this conversation besides in vitro models: her unique perspective as both patient advocate and researcher. Her vantage point has helped her see what her colleagues miss—for instance, the way medical language contributes to keeping endometriosis siloed and overlooked. In 2009, she started noticing something in her lab: physicians like Dr. Isaacson often referred to her disease as "benign." She knew what they meant, of course: non-cancerous. But the word made her wince. As a patient, it felt dismissive. More importantly, the word was signaling the wrong thing to funding entities like the NIH. "If I write that I'm studying a benign disease, who's going to give me money?" she says. "It's not a disease. It's just like: live with it." In 2019, she began campaigning to remove the word "benign" from endometriosis research. Today it has virtually disappeared from conferences and medical papers in the field.*

---

* In case anyone forgets, she reminds them in her email signature: *Please don't refer to endometriosis, adenomyosis, or fibroids as "benign disease"—they are not benign, they are "common and morbid."*

Once, Griffith thought of endometriosis as her cross to bear. Today, she is embracing both halves of her identity to bridge the divide between medicine and the public. "Everybody has their little piece of the puzzle," she said one evening at her home in Cambridge. "It's all a giant mosaic. We put our tiles in, and the picture emerges."

⌣

There is a second meaning to the idea that endometriosis is more than a "women's disease." It isn't just about the disease physically reaching beyond the uterus. It's the fact that endometriosis is far from a disease of neurotic white women, as it was thought of in the days when Griffith was diagnosed.

In 2019, Maisha Johnson was in bed with a heating pad on her midriff. Scrolling through Facebook, she came across a video in which actress Tia Mowry talked about living with endometriosis, and mentioned that the first doctor to take her condition seriously was Black. Maisha was thrilled to see another Black woman sharing her story—but then she scrolled down to the comments. White women, she recalls, were commenting that stories like these create division within the endometriosis community. Many were saying, essentially: "Why do you have to make it about race? Endo affects all of us the same way!"

Maisha was frustrated that those who see the systemic problem of gender bias within medicine couldn't recognize that racial bias, too, is systemic and often compounds the problem. In her own journey as a Black woman with endo, she recalls instances where doctors assumed she was seeking opiates, and instances where her pain was ignored. "If I'm seen as a woman who's prone to hysteria, and then also a Black woman who's not as affected by pain, then obviously if I'm talking about being in pain, then I'm just exaggerating and it can't possibly be that bad," says Maisha, a thirty-four-year-old content writer for *Healthline Media* who writes about chronic conditions and mental health. This is particularly vexing when it comes to a disease that requires surgical evidence to get a definitive diagnosis—yet doctors are often reluctant or unwilling to provide that surgery. Seeing those comments motivated her to write a blog post on her experience with race and endometriosis for *Healthline*.

Jaipreet Virdi, who is deaf and Southeast Asian, told me about a similar experience. Virdi was thirty-five and working as an adjunct history professor of medicine in Toronto, researching disability and gender, when she felt what she would later learn was a nine-inch mass in her abdomen. Time after time, her husband would accompany her to the ER and she would spend hours in the waiting room screaming in pain, only to be sent home. "After the third time it became really apparent they were looking at me like I was a drug-seeker," she says. That was the first instance she experienced "very subtle racism, class bias as well" in a medical setting. Eventually, she began demanding that her medical files reflect that she was a historian of medicine, so that doctors would respect her expertise.

Only after three more emergency room visits did a doctor actually feel her mass and investigate further. All of the doctors involved in her eventual diagnosis, she remembers, were people of color. She was finally sent in for surgery, where surgeons tried to remove as much endometrial tissue as possible from her ovaries, intestines, and bladder. Now forty, Jaipreet has come to terms with the fact that children may not be in her future. What frustrates her is that if doctors had taken her pain seriously earlier—when she had painful periods, when she was fainting in the bathroom—she may have had options.

Finally, associating endometriosis only with those who look stereotypically feminine means that many LGBTQ people—particularly trans men, masculine-presenting women, and nonbinary people—have an even harder time getting doctors to recognize and treat their disease. Gender-diverse patients report that doctors often lack knowledge of how hormones or other treatments can affect the disease in different bodies. "Many clinicians are unable to disentangle gender from anatomy when it comes to providing care," Dr. Frances Grimstad, a gynecologist at Boston Children's Hospital, told *VICE* in 2020. Worse, doctors often lack basic respect for their gender identity, leading them to feel discomfort around sharing their true concerns.

That was the experience of Cori Smith, a twenty-eight-year-old trans man from Rochester, New York, who has endometriosis. Cori's first period, at thirteen, was excruciating. Six months later, he was in the emergency room for a burst ovarian cyst. After being diagnosed with endometriosis at seventeen, he underwent several surgeries to remove endometrial tissue. At the same time, he was figuring out his gender identity. From an early age, he felt sure that he was a boy; at age twelve or thirteen he saw the word "transgender" on the cover of *People* magazine, and knew that was what he was. Still, he put off transitioning because he was constantly dealing with his health problems. When he was twenty-two, he finally went on testosterone and got top surgery to remove his breast tissue. Eventually, knowing he wanted to have biological children someday, he froze his eggs and had his ovaries and uterus removed due to complications with his disease.

Despite his lower levels of estrogen and lack of female reproductive organs, his disease returned, baffling his doctors. "As a girl, they thought I was just a hormonal teenager that just wanted attention," he said in a video interview with *NowThis News* in 2018. "No matter which version of my life, they still ignored it." His experience echoes that of thousands of women who are prescribed hysterectomies to deal with their disease, yet continue to suffer. Six years after transitioning, Cori still finds himself in the OB/GYN office more than most women. "Because of all that, I'm more aware and in tune with the problems that women face in the medical system," he says. For him, endometriosis "is so rooted in who I am and what I've been through that for me it's my story. I don't walk away from it, I kind of walk towards it."

After sharing his story with *NowThis News,* Cori says he's received hundreds of personal emails and messages from others in the endo community. His experience has led him to believe that the numbers of trans and nonbinary people with endo are likely far higher than documented, and that the compounding barriers of having a stigmatized "women's disease" and not identifying as a woman conspire to keep them silent. Continuing to speak about endometriosis as a "women's disease," to him, obscures the reality of this condition, and makes the trans and nonbinary people who suffer invisible. "Breast cancer happens to men," says Cori. "I just don't think it should be gendered."

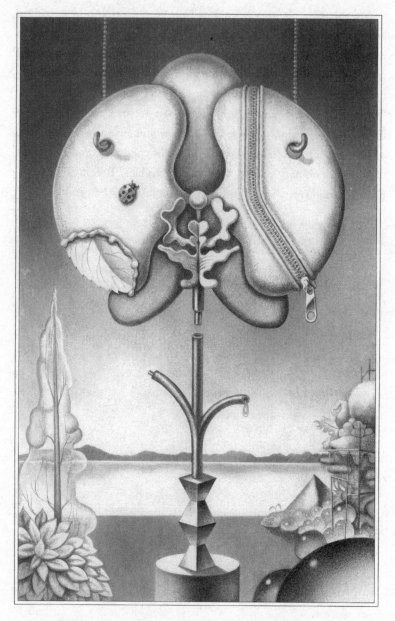

What I've come to realize is that everything a man has, a woman
has; everything a woman has, a man has, anatomically.

—DR. MARCI BOWERS

CHAPTER 8

Beauty

(NEOVAGINA)

D r. Marci Bowers stood before the abyss. "You don't want a vagina like this," she said, igniting titters in the windowless room of thirty-odd conference-goers. "My sense is that it was a cavern." On the screen behind her was a red desert cliff with a round hole cut clear through to the sky. This, she said, was the result you could expect if you were a male-to-female patient going to an American surgeon in the 1960s or '70s. At that point, things like pleasure, sensitivity, or an aesthetically pleasing exterior were not a priority. "If you could make a hole, that was good enough," she said.

Bowers pressed her hands together in front of her. "Many of the post-ops were fine if the legs were mostly together, but when you parted the legs you had kind of an appearance like that"—she spread her hands apart, and pointed to the gaping hole in the rock—"and sometimes you even got an echo!"

Her next few slides were photographs of her own patients after their male-to-female operations. First, a vagina and vulva immediately post-op. Next, a close-up view of the same patient after they had healed fully. Click: another vagina. The genitals onscreen were virtually indistinguishable from any other, except for two faint purple scars running along the labia majora. Under a tuft of pubic hair, both sets of labia and a pink, hooded clitoris were all clearly distinguishable. "This is another good example: minimal scarring, symmetry, labia minora," Bowers said. "Pretty cool—it's amazing that we can do it." The difference was clear: A hole was no longer good enough.

It was 2007, and Bowers was at a conference in Tucson, Arizona, put on by the International Foundation for Gender Education to speak on the past and future of gender affirmation surgery—or, as she called it at the time, genital reassignment surgery. "I think that's the most appropriate name for it," she said, shoulder-length blond hair cascading over the rims of her glasses. "We're not changing gender. Gender is what's there. It's been with us since age three or four years old. We're just simply aligning the genitalia with the gender."*

The room broke into applause.

When Bowers started doing this surgery back in 2003, what struck her was that men and women have all the same external parts, just arranged slightly differently. "The fact that we deconstruct and then reconstruct genitalia, you realize, for one, that male bodies and female bodies really are just the same," she says. "It really is a continuum." After more than a decade of working as a gynecologist and obstetrician, she knows intimately that genitalia comes from the same structures, the same tissues. That's why, for her, turning penises and testicles into vaginas and clitorises—and, less commonly, vaginas and vulvas into penises and testicles—makes perfect sense.

As the first (known) transgender woman to become a gender affirmation surgeon herself, Bowers brings a unique perspective to her work. Her goal is to give patients what she wanted when she got her own male-to-female surgery in Mexico, back in 1997. "Had I been

---

* Today, the accepted term is "gender affirmation surgery," to indicate that the surgery changes genitals but reaffirms a person's existing gender identity. As Bowers puts it: "Sex is between the legs. Gender's between the ears."

going through the process at a later date I would have wanted something that looked better, functioned better," she says. "Especially with my background in gynecology, I really felt like what I'd seen so far was fairly limited—fairly lacking." Since 2003, she has been pushing the boundaries of what surgeons can do to give patients the appearance and sensation they desire.

In her field, she has earned a reputation for her careful work, artistry, and focus on female sensitivity and pleasure—qualities that once earned her the nickname "the Georgia O'Keeffe of genitalia." Her waiting list for surgery is five years long and growing. "It's not just a hole for a man to ejaculate into," she says today. "It has sensitivity, it has shape, it has beauty. The labia should frame the vagina. It's not just separate anatomical structures like clitoris, urethra, vagina, sort of separate, floating in the air. They need to have a relationship with one another."

It's a gynecological perspective—a woman's perspective.

—

Born in Oak Park, Illinois, in 1958, Bowers remembers having crossgender feelings by the time she was four or five. "I was never a boy's boy," she said in a 2007 documentary about her work, *Trinidad*. "I was beaten down because I was a skinny kid and effeminate." She felt ashamed of her feelings and pushed them down, playing with trucks and action figures instead of dolls. When she went to medical school at the University of Minnesota, she decided to specialize in obstetrics and gynecology. After completing her residency at the University of Washington in Seattle, she rose to become the head of the OB/GYN department at the Swedish Medical Center/Providence.

Bowers hoped having a career in women's health would satisfy the feelings she had. At her practice in Seattle, she delivered more than three thousand babies, performed gynecological surgeries, and occasionally provided aftercare for trans patients. She learned intimately the relationships between the parts of female anatomy—the way the labia folds over and covers the clitoris, the position of the urethra in relation to the clitoris. She fell in love with a woman, married, and became a parent to two daughters. But as the years passed, her conviction only grew stronger. "The more I felt a part of their lives, the more I felt that, actually, this should be *my* life," she said.

By the time her wife was pregnant with their third child, a boy, she could hardly look in the mirror. In 1995, she decided to go on feminizing hormones. When their son, Thomas, was born, her transition was well under way. He's never known her as anything but Marci. Twenty-two years into her marriage, she underwent gender affirmation surgery. She and her wife remained legally married, but Bowers says she considers their relationship more of a "sisterhood."

Given her journey, Bowers understands her patients better than most. "What patients really experience in terms of their emotion after surgery is a sense of relief," she says. "They feel like they're one with their soul, finally." Yet when asked whether her own journey makes her better suited to this work, she prefers to point to her extensive experience as a gynecologist and obstetrician. "That's much more important than my trans history," she said in *Trinidad*. "That's just an interesting artifact. But the fact is that first I'm an artist, secondly I'm a surgeon, third I'm a gynecologist, and about eighth I'm a trans person."

Her skills have attracted notice outside the transgender community as well. In 2007, Bowers got a call from the director of Clitoraid, a nonprofit organization that trained volunteer surgeons to provide clitoral reconstruction surgeries to women who have been scarred by genital cutting. Given her expertise in constructing and reconstructing genitalia, they wanted to recruit her as their head surgeon. For Bowers, it was a no-brainer. She traveled to Paris to learn the technique from Dr. Pierre Foldès, and went on to train more than twenty surgeons herself. She quickly learned that the clitoris was much larger than she had appreciated, cementing her understanding of how similar male and female sexual organs are.

To demonstrate, she takes a hot-pink anatomical model of a clitoris out of the pen cup on her desk. It's August 2019, and we're sitting in her consulting offices in Burlingame, California; Bowers has just gotten back from her annual trip to Nairobi, Kenya, where she performed clitoral restoration on sixty-five women. The model is immediately familiar: it looks just like the one on Dr. Ghada Hatem's desk. Like Hatem, Bowers uses it to show her cisgender patients who have undergone genital cutting their true anatomy and impress upon them how much untouched erectile tissue they still possess. "Every cisgender woman on the planet has about three to five centimeters of clitoris in front of the pubic bone," she tells me. In most forms of female genital cutting, "they remove less than 3 percent."

Bowers draws a direct parallel between her work with these patients and her gender affirmation surgeries. "In my life with the transgender community, and the life I lead with these women, I offer the one thing that I think should sustain us," she said in a 2017 TEDx talk, "and that is hope."

———

The first time Roxanne Euber flew in a plane was June 2019. It was the first day of Pride Month, and she was flying with her girlfriend, Elle, from Denver to San Francisco to get her gender affirmation surgery with Bowers. Normally Roxanne, a forty-six-year-old property manager from Colorado Springs with long auburn hair and wire-rim glasses, drives everywhere; she likes the control she has behind the wheel. But when she felt the plane accelerate and take off—the drop in her stomach, the feeling of weightlessness—she let go. After three years of waiting, she was ready to give herself over to Bowers's scalpel.

"This is my one shot," she said. "This is everything to me. This is my chance to be a whole person."

Two days later, in her pre-surgery consultation in Burlingame, Bowers asked Roxanne when she first felt she had been assigned the wrong gender. Roxanne recalled traipsing around her childhood home

wearing her mother's heels and trying on her Avon lipstick samples
when she was four or five. Her grandfather, a WWII veteran, soon put
an end to that. "I had my identity beaten out of me at a young age," she
told me later. "It became so buried for so long. They told me what I was
supposed to be, and that's what I tried to be." For decades she lived as a
man, with a bushy brown mustache and a love of Shaolin martial arts.
At twenty-five she married a woman, settled down, and had a child.

It never felt right. At forty, it clicked: she was a woman. "I had no
idea what I was going to look like when I started this process, and I
didn't care," she says. "I was like, I don't know if I'm going to be pretty,
I don't know what I'm going to look like, but I will be me. And that's
all that matters." She got on hormones, a combination of estradiol (a
form of estrogen also used in menopausal hormone therapy), and spi-
ronolactone (an androgen blocker that reduces body hair and contributes to breast growth). She changed her wardrobe from baggy jeans
and T-shirts to flowing skirts and blouses. She began dating men—and
later, one woman, Elle. One day she looked in the mirror, and finally
recognized herself.

Roxanne has gender dysphoria, which the *Diagnostic and Statistical
Manual of Mental Disorders* (DSM-5) defines as "the distress that may
accompany the incongruence between one's experienced or expressed
gender and one's assigned gender."* Yet that bloodless definition does
little to get across what it feels like to live with it. Roxanne compares
it to having a toothache that never goes away. Sometimes it fades into
the background, a dull ache that makes her feel a little off. Other times
she's painfully aware of the discomfort. When she changes into pajamas
at night, for instance. Any time she gets sexually intimate. When she's
in the shower and accidentally looks down at herself.

---

* This term "gender dysphoria" has been in use since 2013, replacing the outdated
"gender identity disorder." However, it is still debated whether "dysphoria" also
pathologizes transgender people as medically deviant and in need of fixing.

"And how would you describe your gender identity now?" asked Bowers, who was typing notes on a black tablet.

Roxanne didn't hesitate. "I'm 100 percent a woman."

She folded her hands in her lap and looked down at her nails. They were almond-shaped and painted a shimmering peach, except for the ones on her ring fingers, which were burgundy. When she first started transitioning, her nails were the main way she expressed herself. At that point she was still presenting as a man at work, still getting called by her deadname (the name given to her at birth, which she does not use), which felt like a punch to the gut. But no matter how bad things got, she could look down and see proof that she controlled her identity. "They're still that anchor," she says. Today she calls them "trash," hardly up to her standards, but it doesn't bother her as much. It's no longer her only outlet of expression.

"Just 'woman'?" asked Bowers, looking up from her screen.

"Yes. Just the person I should have been."

—

On the morning of her surgery, three days later, Roxanne pumped her fist in the air in triumph. It was seven thirty a.m., and she was standing outside the Mills–Peninsula Hospital, a shimmering façade of blue and silver windows that glinted in the morning sun. "We're here," she whispered.

Roxanne never fully believed that this moment would come. Besides the three-year wait, finances were a huge barrier. Fortunately, her health insurance covered the surgery itself. But she still had to pay a $2,500 deductible, airfare, food, copays, and the Airbnb while she recovered. Thousands of dollars' worth of laser hair removal and electrolysis had whittled down her savings to almost nothing. Elle, a mechanic, had been working at two different bike shops since January to make sure this trip could happen. Plus, they still had to save up enough money for Elle, who is also trans, to get the same surgery.

Not every transgender person needs or desires genital surgery to feel at home in their their body. But for Roxanne, it felt like a final step in her transition. "If I could just take some magic pill and have it all go away, I would," she said in the car ride over. "I wouldn't wish this hell on anybody. But we are who we are. We cannot change who we are. All we can do is change the flesh that we occupy to reflect our actual selves."

Just before surgery, Bowers opened the curtain to the pre-op room and strode in, looking ready for summer in a sundress and blazer. Roxanne was already wearing a blue-and-white-striped hospital gown, an IV attached to her forearm. Elle was sitting beside her. Bowers took one of Roxanne's hands, which were clasped across her chest.

"It's going to go beautiful," she told her, her voice radiating confidence. "We're going to make something pretty out of this."

Roxanne stamped her feet under the blanket and squealed with joy.

"I mean, it's going to hurt," Bowers clarified. "You'll have a lot of pain, and you'll have all the dilation, and all those things as you go through. But it's going to be fine."

Roxanne nodded. "It's all part of the journey."

"All part of the journey," Bowers agreed.

———

When Bowers steps into the operating room, everything else fades away. All she sees is the anesthetized patient before her—today, Roxanne—and the potential to transform her genitalia into something beautiful. Before her is an unnecessarily large protuberance—"The penis is just a large clitoris. In fact, I don't know why they don't just call it a large clitoris," she likes to say—that she can peel back, pull apart, and reshape into a thing of grace: a vagina.

After marking up her canvas, she makes an incision into the scro-

tum and removes the testicles, a procedure known as an orchiectomy.*
She saves the scrotal skin, wrapping it in moist gauze. Next she peels
back the skin of the penis, removing much of the tissue but taking care
to preserve the sensitive head, or glans. (The remaining foreskin will
become the hood of the clitoris.) She separates out the spongy urethra
and shortens it, repositioning it alongside the new vaginal canal. To
make the canal, Bowers turns the penis inside out like a stocking, pushing the hollow tube of skin back into the pelvic cavity. This skin will
be become the lining for the inside of the neovagina—the term those
in the medical field use for surgically crafted vaginas.

Finally, she crafts the inner and outer labia from the saved scrotum and carefully sews them together. Within a year, Roxanne's scars
should fade away almost completely.

Genital surgery is often portrayed as a full transformation, from one
thing into its opposite. Yet the surgery relies on the fact that male and
female genitalia are very much the same. Within the penis are three
columns of erectile tissue, called *corpora*. The two larger columns, called
corpora cavernosa (cavernous bodies), surround the urethra and extend to
form the glans penis. The smaller, flattened mass that lines the sensitive underside of the penis is called the *corpus spongiosum* (spongy body).
During erection these bodies swell with blood, growing to contain
almost 90 percent of the organ's blood supply. These tissues correspond
perfectly to the arms and the bulbs of the internal clitoris.

Bowers harvests these erectile columns and uses them to craft a clitoris, with the head of the penis becoming the new glans clitoris. This
new clitoris will have full sexual sensation. It can take up to a year, but
eventually, patients like Roxanne will be able to orgasm. "Generally,

---

* If that word sounds familiar, you're right. Orchids got their name because their
tuberous roots were thought to resemble the male genitals. Both terms derive from
the Greek word *orchis*, which means testicle. The word is also the root of *cryptorchidism*,
the condition of undescended testicles.

if you were able to before, you will afterwards," Bowers says, adding, "whatever your relationship—boys or toys."

Getting the dissection right, putting the parts in the right place, avoiding complications—all that is routine. "The most difficult part is making it look perfect," Bowers says. For her, this surgery is a high art, demanding the aesthetic touch and attention to detail of a sculptor. Every vagina she creates, she wants to have the elements of "balance, definition, grace, delicateness." Every operation, she stresses herself out trying to create a body part that her patient will be pleased with for the rest of their life. Every time she finishes, it still feels like she's accomplished a miracle.

"And then it comes out perfect," she says.

———

In a sense, Bowers is re-creating the female anatomy. But she's also redefining it, says Eric Plemons, a medical anthropologist at the University of Arizona who studies genital surgeries.

A rhinoplasty surgeon, Plemons points out, might change the shape and size of the nose, shaving off cartilage to better align it with current beauty ideals. Each nose surgeon may have a different style; they may prefer to create a slightly straighter or smaller nose. "But the question is not 'What is a nose?'" Plemons says. For genital surgeons, "there's still an ongoing competition over authoritative representation—who gets to say what this thing is, and what it should do, and what it should be like, and what it should be for." Surgeons like Bowers help produce what is considered "normal," and even help set the standard for what a vagina should look and feel like.

As a result, almost every surgeon does this procedure differently. Some focus more on aesthetics, some on pleasure. Their techniques are so distinct that most surgeons can tell just by looking at a postoperative patient exactly who did their surgery. "Every surgeon has his or her own twist," says Dr. Pierre Brassard, a Montreal-based gender affirmation surgeon. "Each surgeon is using the same parts, but differently."

Brassard, who performs hundreds of vaginoplasties a year, wouldn't tell me exactly what makes his own operation special. "We want to keep the Montreal edge," he demurred.

Belgian surgeons, for instance, says Plemons, have developed the "gold standard" of phalloplasty: a smaller, more modest appendage that allows the wielder to urinate standing up—a key mark of masculinity for many patients—and can be outfitted with an erectile prosthetic. American surgeons joke that "the Belgians make a very nice penis if you want something that small," says Plemons. On the other hand, Belgians point out that American surgeons are obsessed with making an "overlarge penis" that is unwieldy to manage. "If you think it's supposed to be this massive thing that people just gasp over, you can make that," Plemons says. "But then it also becomes a very difficult thing to actually use, depending on how you want to use it."

Neovaginas, meanwhile, have historically been constructed around penetrative intercourse. "There's the fantasy of the vagina that can 'take it all,' so it needs to be really big and really stretchy," says Plemons. The most common boast Plemons hears from surgeons is that their work is so realistic that your husband can have sex with you and never know—which helps explain why early vaginas were not exactly pretty. After all, if the main role is to give pleasure to someone else, "Who cares what it looks like?" Plemons says. "It could look like the Fleshlight things you buy from porn stars."

By contrast, surgeons like Bowers aim to craft vaginas that are not only aesthetically pleasing but fully functional. Like a biological vagina, a neovagina must defend itself, clean itself, get aroused, and, perhaps most importantly, give its owner pleasure.

For surgeons, this creates new challenges. Take lubrication: Under the right conditions, the vagina makes its own lube. When a woman is aroused, blood vessels dilate to pump blood down to her genitals, particularly the clitoris and labia minora. Some of these vessels are "leaky," transferring fluid through the walls into spaces between cells of the vaginal lining in a process known as transudation. This creates the "wetness" of arousal. Penile skin doesn't have this quality, so Bowers and other surgeons have developed an alternative surgical technique: they can use a loop of peritoneum, the mucosal lining that surrounds the pelvic organs and one of the few other places in the body that self-lubricates.

It is also crucial to create a canal that has depth and stretchiness. After her surgery, Roxanne will use a series of dilators—rigid, tube-shaped devices that helps the neovagina retain its depth and width—for around twenty minutes a day. Even after she fully recovers, she'll use them twice a week for the rest of her life. Her neovagina will also develop its own microbiome, a protective mixture of microbes that begin settling in as soon as she removes her bandages. Even less is known about these microbiomes than those of cisgender women. But researchers have found that, in a penile-skin-lined vagina, the types of bacteria are more similar to those of the skin and intestines. Similarly, a vagina constructed from other organs will mimic the microbiomes of those environments.*

Even the fact that today's gender affirmation surgeons are considering factors like pleasure, wetness, and protection shows how far this surgery has come. What's clear is that the early developers of this surgery had a very different ideal in mind than Bowers does. For them, it was good enough to remove the penis and testicles—the most obvious outward sign of masculinity, and the glands that produced testosterone—and create a hole for heterosexual intercourse. It took a long, winding journey to get to a place where genital surgery was seen as a way for trans women to experience their own bodies and sexuality.

---

* Studies suggest that this microbial mixture means a patient won't get as many yeast infections and won't be as susceptible to STDs like gonorrhea or chlamydia.

On December 1, 1952, the air crackled with the power of science and technology to better human life and destroy it. Penicillin had just become widespread. Jonas Salk was close to perfecting the polio vaccine. The United States had just conducted its first nuclear test of a hydrogen bomb, thought to be the most powerful weapon known to man. There were rumors America would soon put a man on the moon. And yet, readers of the *New York Daily News* that day saw none of these achievements. Picking up their newspaper from their doorstep, they would have seen only the front-page headline, in capital letters: "EX-GI BECOMES BLONDE BEAUTY."

Two side-by-side photos revealed a gangly young soldier with prominent ears, his hat perched jauntily to the left, next to a stunning woman in profile, her flaxen hair in a chignon, eyebrows arched, a straight Grecian nose. "A World of a Difference," the article announced. It told the story of how Christine Jorgensen, a twenty-six-year-old American veteran with Danish parents, had traveled to Denmark to undergo "a rare and complicated treatment." Science, it seemed, had finally triumphed over nature. "Impossible!" Jorgensen would write in her memoir. "That word was a challenge to me. How did anyone dare say it in the Atomic Age?"

Jorgensen was born in 1926 in the Bronx as a boy named George. Painfully shy and deeply insecure—"a frail, tow-headed, introvert child," she would write in her autobiography—she waffled over her future, at turns pursuing photography and dental school. She finally joined the war effort out of a desire to belong to something bigger. But by the time she was accepted, the war had been won, and it was her task to help bring the troops home.

America was in an age of anxiety. During World War II, women had left the home in droves to pick up men's work in factories, aiding the war effort on the home front. As men trickled back home, those women were unceremoniously ushered back into the domestic sphere. A decade earlier, scientists like Eugen Steinach had planted the idea that men and women were not so different after all; just as light is both a particle and wave, each person contained elements of both. Now the cultural tide had shifted. Women needed to be feminine, and men needed to be masculine. Homosexuality was cause for panic.

Jorgensen didn't feel homosexual; her Christian upbringing had instilled in her a deep moral repugnance for that word. But she knew she didn't belong in her male gender role. After her brief military stint, she returned home to attend photography school in New Haven, Connecticut. In October 1948, she read in the headlines that scientists had used hormones to change a chicken from male to female, and female to male. For the first time, she felt a sliver of hope. What if, instead of trying to act more masculine, she tried to become more feminine?

Soon she was spending hours at the library of the New York Academy of Medicine, reading literature on rare glandular disorders and "sex conversions." She discovered a slim book called *The Male Hormone* by Paul de Kruif, which equated hormonal secretions with manhood and womanhood. To de Kruif, everything boiled down to chemistry. "Chemically, all of us are both man and woman because our bodies make both male and female hormones, and primarily it's an excess of testosterone that makes us men, or an excess of female hormones that makes us women," he wrote. "The chemical difference between testosterone and estradiol is merely a matter of four atoms of hydrogen and one atom of carbon."

Jorgensen was struck by the minuteness of the difference. "At that moment it seemed possible to me that I was holding salvation in my hands: the science of body chemistry," she wrote. After graduating in 1949, she walked into a drugstore and convinced a pharmacist to give her one hundred tablets of estradiol. NOT TO BE TAKEN WITHOUT THE ADVICE OF A DOCTOR, the label read. For years, she took one tablet every week in secret, hoping it would transform her body and psyche.

Finding an American doctor who saw things her way was a challenge. Finally, she learned of the possibility of an experimental "sex conversion" in Europe, and set off on a ten-day boat trip with the last of her savings. She found a welcoming surgeon in Copenhagen:

Dr. Christian Hamburger, an endocrinologist and acolyte of Steinach's work on glands and sexuality. He listened to her explain her life and her desire for surgery. "I think the trouble is very deep-rooted in the cells of your body," he told her. "Your body and all of your body cells, including your brain cells, may be female."

Hamburger told her that it might be possible to suppress her maleness with female hormones, and then surgically transform her genitalia. He would even do the experimental treatment for free—providing she agreed to let him write up her case as the first of its kind. Jorgensen only heard one thing: it might be possible. She didn't hesitate. In 1950, she began taking high-potency estrogen tablets under Hamburger's supervision in preparation for the surgery. On the pills, she felt calmer. Her breasts swelled, and her skin cleared up. She saved all her urine for analysis, which she kept in a two-quart bottle (she called it her *yor mor taske*, which means "midwife's bag" in Danish) and hid it in a black bag.

In September 1951, after receiving an official "permission to castrate" permit from the Ministry of Justice, her Danish surgeons performed the long-awaited surgery. In less than an hour, they removed her testicles, her body's main source of testosterone. But according to her doctors, "the patient had one final ardent wish: to have the last visible remains of the detested masculinity removed." One year later, on November 20, 1952, they performed a penectomy, or removal of her penis, surgically reshaping her remaining scrotal skin into "labia-like formations." From the surgeon's point of view, Jorgensen now had achieved "a harmonious balance between soma and psyche."

In 1954, she would be able to achieve the third and final stage of her transformation in New Jersey. Few details are recorded about exactly what was done, but in her autobiography, she called it an "extremely complicated operation" that took seven hours to perform. "With skin grafts taken from the upper thighs, plastic surgery constructed a vaginal canal and external female genitalia," she wrote. "At that point, I felt at last that I'd completed the transition to womanhood, and except for the inability to bear children, was as complete a person as I'd dreamed of being."

In 1951, Jorgensen wrote to her friends in the States that the "shy, miserable person" she was when she left America was no more. She chose her new name in honor of the physician who had been willing to take her on: Christine.

After her second operation, she wrote a letter home to her parents in New York, who knew only that she had left the country to visit Danish relatives. "We humans are perhaps the greatest chemical reaction in the world, and therefore it is not strange that we are subject to so very many physical ailments. Among the greatest working parts of our bodies are the glands. Several small, seemingly unimportant glands, and yet our whole body is governed by them. An imbalance in the glandular system puts the body under a strain, in an effort to adjust that imbalance," she wrote, and she "had such an imbalance." She enclosed a few photos she had taken of herself earlier that year. "I am still the same old 'Brud,'" she wrote, using her childhood nickname. "But Nature made a mistake, which I have had corrected, and I am now your daughter."

Weeks later, the reply came by cable: LETTER AND PICTURES RECEIVED. WE LOVE YOU MORE THAN EVER. MOM AND DAD.

Jorgensen's newfound peace would not last long. An American reporter had intercepted a copy of her letters to her parents and was about to turn them into a sensational news story. At New York International Airport (now John F. Kennedy International Airport), she walked off the plane ramp and into a chaos of flashbulbs and shouting reporters. Dressed in pearl earrings and a heavy mink coat, teetering on heels, she was cast by the press as a Marilyn-esque star: aloof, classy, and very heterosexual. Here was a person who had gone from the height of masculinity—a military man, as the press described her, a bit misleadingly—into a "blond bombshell." In a moment, her private journey became a public symbol for the world.

Jorgensen didn't want to be an icon. "I merely wanted to correct

what I considered to be a misjudgment of Nature, so that I might physically and legally become the person I felt I was intended to be," she
wrote. But she began to see the larger injustice in the medical establishment's refusal to help those like her. Soon she was receiving thousands
of letters from individuals desperate to get the same treatment she had.
"Can you realize what success for me will mean to literally thousands
of people?" she wrote home to friends in New Jersey. "It may mean
new hope and life to so many people."

—

One of the letters was from someone named Dr. Harry Benjamin. An
endocrinologist and sexologist born and trained in Germany, Benjamin
was a disciple of Dr. Eugen Steinach, who was spreading his glandular
gospel in New York. After Benjamin's colleague Alfred Kinsey introduced him to his first "transsexual" patient in 1948, he had grown
increasingly fascinated by those who felt they had been born into the
wrong sex.* In the early half of the century, this condition was little
understood and rarely written about; when it was, it was usually associated with homosexuality and cross-dressing.

Benjamin, who would become known as the father of transgender
medicine, saw it as his mission to explain an oppressed and misunderstood
minority to American clinicians. "To me it is just a matter of relieving
human suffering the best way we can," he told *Esquire* magazine in 1967.
Transfixed by Jorgensen's story and the courage of her Danish doctors, he
wrote to her to offer his services. Benjamin, who had received his share
of letters from desperate patients seeking treatment, offered to compose

---

* At the time, the term "transsexual" was used to refer to those who felt themselves
to belong to the opposite gender. "Transvestism," by contrast, referred to someone
who dressed as the opposite gender. In the 1980s, "transsexual" was replaced by the
term "transgender." Today, both terms (transsexual and transvestite) are considered
outdated.

a form letter for her. Jorgensen began rerouting her letters to him—first dozens, then hundreds, then thousands.

Jorgensen was equally moved by Benjamin's struggle against the legal and medical hurdles that transsexuals faced. "What I'm trying to do is not as great as the big medical discoverers in the past, but it will be a contribution," she wrote to him in 1953. "With God's help and those few who believe as you do, I know this will be a step into the future understanding of the human race." For Benjamin, Jorgensen was a particularly fascinating case; she would be the subject of many of his future medical papers. In a letter to another patient and longtime friend, he mused that. "Christine's 'transvestism' was only a part, the external or symbolic part of her problem. The urge goes much deeper. We lack a proper scientific term for it. I would describe it as an 'obsessive urge to belong to the opposite sex.'"

Benjamin was born and trained in Berlin, where gender affirmation surgeries were performed until the Nazis came to power in 1933. In America, he fell in with a group of sexual reformers, including Margaret Sanger and Kinsey, who crusaded for everything from contraception to liberalized divorce laws. The American medical system, in Benjamin's eyes, was decades behind the "more enlightened attitudes" of countries like Denmark, Holland, and Sweden. He lamented that Americans had to travel to Africa, Asia, or Europe to get genital surgery, and compared these surgeries to nose jobs in Germany: Fifty years earlier, he pointed out, plastic surgeons in Germany who did rhinoplasties were branded as quacks; now it was routine.

For Benjamin, sex was like a pointillism painting—the closer you leaned in, the murkier it became. "Instead of the conventional two sexes with their anatomical differences, there may be up to ten or more separate concepts and manifestations of sex and each could be of vital importance to the individual," he wrote in his 1966 book, *The Transsexual Phenomenon*. These manifestations included chromosomal sex, gonadal sex, genetic sex, social sex, psychic sex, and more. Most people, he said, experienced a "symphony of the sexes": the harmony that comes when each element of your gender identity aligns. "Transsexuals," by contrast, experienced a profound disharmony between their "psychic sex" and their bodily sex.

As a gland man, Benjamin agreed with Jorgensen that her problem was biological. Disdaining psychoanalysis as a "cure," he believed instead that scientists would soon discover a biological cause—perhaps a genetic factor, or the influence of hormones during fetal development—for her condition. But whatever the cause, in his mind, there was only one way to help these individuals find harmony between their psychic sex and their biological sex: complete gender transformation, often including surgery. "And so I ask myself, in mercy, or in common sense," he wrote, "if we cannot alter the conviction to fit the body, should we not, in certain circumstances, alter the body to fit the conviction?"

For the vast majority of his patients, though, Benjamin could offer only sympathy, a diagnosis, and a prescription for (usually) feminizing hormones.* For many years, he was the only physician in the United States to provide estrogen, in the form of injections and the early birth-control pill Enovid. (The idea was not initially his: In the 1920s, a transgender patient asked Benjamin for a trial of the recently developed Progynon Depot. He agreed and, after witnessing relief from her gender dysphoria, came to recommend hormones to many of his patients. He found that estrogen calmed the patient, lowered libido, and led to breast growth and other desirable feminine traits.)

Nevertheless, Benjamin and Jorgensen offered something essential to thousands of Americans who longed to transform their sex: hope. In addition to becoming a symbol of the limitless power of science, Jor-

---

* Most of Benjamin's patients were assigned male at birth. However, one of his female-to-male patients, the philanthropist Reed Erickson, would become a major financial supporter. Erickson created the foundation that paid for Benjamin's research and helped fund the Johns Hopkins program.

gensen had captured the imagination of those who had found themselves closed off from every avenue. Her journey set the blueprint for thousands of others who would come to envision the quest for hormones and surgery as the realization of their dream of living wholly in their felt gender.

⌣

And yet few could follow where Jorgensen had led. In 1953, overwhelmed by demand, Denmark turned down more than four hundred appeals for transformation operations, many of them from Americans. Nor could American surgery-seekers find a willing surgeon in their own country, unless they could prove that they were intersex patients, born with atypical genitalia.

Up until the 1960s, it was virtually impossible to get gender affirmation surgery in the United States. First, many surgeons considered the operation dangerous and morally questionable. But even if they wanted to perform it, they were stymied by an obscure group of laws known as the mayhem statutes. These statutes dated back to the days of Henry VIII, when some British youths seeking to avoid conscription went to such lengths to as to ask their doctors to remove healthy parts of their body needed for combat—fingers, toes, even the occasional hand or foot. To prevent this, the king enacted a ban on removing any part of a soldier that was necessary for his defense.

Though no doctor was ever actually prosecuted for mayhem, fear of reprisal effectively squelched the procedure in the United States. Benjamin railed loudly against what he saw as an inhumane and morally backward statute. "Eventually a Supreme Court decision may be required to ban the specter of the mayhem statute for surgeons and allow them to act in accordance with science and their own consciences," he wrote. Still, the most he could do was continue to offer his patients hormone treatment, and refer them to one of the handful of receptive surgeons he knew around the country.

As a result, the few dozen gender affirmation procedures done in midcentury America were risky, dangerous, and done under the cover of strict secrecy. One early patient, a sex worker named Patricia Morgan, was able to find a willing surgeon in Los Angeles through Benjamin. Her first surgery in 1961, she recalled, lasted eight hours. Doctors removed her penis and pushed her testicles up into her abdomen; by not removing them, they hoped they could avoid being prosecuted under the mayhem statute. She awoke to blood, pain, and a tangle of wires, tubes, and catheters. "I was just a glob of aching flesh," she recalled.

By the 1966, these patients would have a new option: the Johns Hopkins Gender Identity Clinic in Baltimore.

In the 1950s, Johns Hopkins had become the country's nucleus for treating intersex children born with atypical genitalia—for instance, a larger-than-usual clitoris, a smaller-than-usual penis (often there is virtually no difference, making the label arbitrary). For clinicians, these children posed a problem. The moment we are born, virtually every person is sorted into a category: boy or girl. The very existence of non-boy, non-girl bodies threatened that simple classification scheme. It challenged the entrenched belief that sex is one unified effort, a symphony in which each instrument is adjusted to the same key—an assumption that Benjamin and others were just beginning to question for their transgender patients.

A New Zealand–born psychologist, John Money, had figured out a way to deal with these intersex children.[*] After writing his Harvard dissertation on the psyches of people born with atypical genitalia, he

---

[*] Today, some in the medical community refer to these children as having "disorders (or differences) of sex development." However, this term has come under scrutiny for the same reason that "gender identity disorder" was eventually deleted from the DSM: the word "disorder" is medicalizing and implies something is wrong with the person. In reality, intersex activists argue, it is the culture, not the body, that needs to be changed.

concluded that successful patients were those raised unequivocally as either boy or girl. The path to becoming a "normal" adult—one who fit into the box of either female or male—was to choose a sex role as early as possible, and stick with it. The question was, what should determine whether a child would be raised as a boy or a girl? For Money, it was not necessarily their chromosomes, hormones, or even gonads. It was primarily the shape of their external genitals, and "how well they lend themselves to surgical reconstruction in conformity with assigned sex."

Money believed that a person's gender role—a term he coined to describe a person's felt sense of themselves as male or female, and would later call gender role/identity—was shaped by their environment, not anatomy. That identity, he believed, solidified after around eighteen months. It was therefore crucial to mold the child to fit their assigned gender as soon as possible, like folding a just-baked fortune cookie before it has a chance to harden. The parents would raise the child in the assigned gender and give additional hormones at puberty to reinforce that role.

In practice, this meant crafting atypical genitals into vaginas and vulvas, which was surgically simpler than the reverse.* Moreover, Money believed that a man with a small penis would face more psychological troubles than a woman with a surgically shaved clitoris. Money's team reduced or amputated the clitoris to create "a clitoris of appropriate size," widened the vagina, and sent the baby home as a girl. The priority was not pleasure and sensation but the ability to have penetrative sex (and if possible, bear children). And yet, Money and his colleagues declared: "Clitoral amputation in patients living as girls does not, so far as our evidence goes, destroy erotic sensitivity and responsiveness, provided the vagina is well developed."

In a way, Money's theories held the same appeal as Sigmund Freud's writings on female sexuality. They offered a seemingly definitive answer—backed by the "scientific" language of psychology—to a

---

* As one prominent urologist put it: "You can dig a hole, but you can't build a pole."

thorny, delicate question. As historian Elizabeth Reis wrote in her 2009 book *Bodies in Doubt: An American History of Intersex*, his theories filled a vacuum of knowledge, convincing perplexed doctors around the country that the riddle of so-called "hermaphroditism" had been solved, and that these children could be raised to become healthy, heterosexual adults. Although he was a psychologist, not a surgeon, his protocol would be the reigning medical paradigm for nearly four decades.

As those children grew up, they would face devastating complications: stabbing pain, urinary problems, a lack of sensation and orgasm, scarring, and for many, a deep, lifelong shame. His most infamous patient was not an intersex child but a boy who had suffered a botched circumcision, named David Reimer. Money's team surgically reshaped his genitals to resemble female ones, and sent him home a girl. After a tumultuous childhood, Reimer learned what had happened to him, and chose to live as a male. At thirty-eight, he took his own life. It would take decades of vocal activism from the intersex community for human rights groups, and some doctors, to recognize the deep and lasting harm these surgeries inflicted on children—an uphill battle still being fought in hospitals and healthcare systems today.

Money framed intersex surgeries as a way for medicine to correct nature's "error"—the same language Jorgensen had used to describe her condition. The key distinction was that, in this case, the infants had no choice in the matter.* Today, many intersex activists hope to bar all medically unnecessary surgery until after puberty, when the patient is old enough to decide what, if anything, they would like done to their body. The main difference between genital surgeries in the transgender community and the intersex community "is like the difference between birth control and sterilization—it lies in the agency of the person undergoing the procedure," says Susan Stryker, a leading scholar of transgender history and professor emerita at the University of Arizona.

Over the next decades, the Hopkins team performed hundreds of

---

* Meanwhile, parents who agreed to surgery were often ill-informed and in shock.

these surgeries, the vast majority of them medically unnecessary, many with disastrous results. Yet despite their very different ends and ethics, these procedures provided the basic surgical framework for a different kind of surgery—one that many in the transgender community were eagerly seeking.

Despite believing that a change in gender role was "psychologically injurious" after the early months of life, Money was an early advocate of gender affirmation surgery. After hearing about Jorgensen's transition, he persuaded Johns Hopkins to offer a similar procedure. In 1966, Hopkins quietly became the first American hospital to open its doors to gender affirmation surgery. As soon as the word was out, surgeons found themselves facing a tsunami of demand. "In a period of two weeks, we got about 3,000 letters from all over America from transsexual patients," said Hopkins' chief of plastic surgery, Dr. Milton Edgerton, in 2016. The letters "were eight to ten pages long; people begging for surgery."

Hopkins would serve as a model for the nation. By the 1970s there were twenty university clinics offering similar surgeries, on campuses ranging from Stanford to Northwestern to the University of Minnesota. During the first two years the Hopkins clinic was open, at least two thousand people applied for surgery, many of them directed by Benjamin. The clinic admitted twenty-four of them. By 1972, out of the five hundred Americans who had undergone gender affirmation surgery in the United States, just thirty-two of them had been at Hopkins. Patients had to jump through stringent hoops, including psychological testing, interviews with family members, IQ tests, and physical exams. They then had to undergo hormone therapy and live in their chosen gender for at least six months before becoming eligible for surgery, which cost $1,500 (around $10,000 today).

As it turned out, Hopkins doctors wanted a very specific kind of patient: one who was already hyperfeminine, who would "pass" effortlessly in their new gender role. Even Benjamin said he wanted to know that "a reasonably successful 'woman' could result." (What made a woman, he once wrote, was that she was "able to function as a female—that is to say, she can have marital sex relations.") It was exactly the result they were hoping for from intersex patients: the ultimate sign of success was when a patient disappeared into society afterward. Dr. John Hoopes, a plastic surgeon and the chair of the clinic, bragged that of the ten patients he had "converted," three were already married and three more were engaged.

Johns Hopkins never wanted to open the floodgates to surgery. In fact, the initial aim of the clinic was experimental. "This program, including the surgery, is investigational," Hoopes told the *New York Times* in 1966. "The most important result of our efforts will be to determine precisely what constitutes a transsexual and what makes him remain that way." In a way, clinics like Johns Hopkins were more about affirming the gender order than questioning it. "Trans is a threat, in a way that has to be contained," says Stryker, the author of *Transgender History: The Roots of Today's Revolution*. "You can totally take apart who's a man or who's a woman—as long as you put them back in column A or column B."

These conservative undertones were not lost on patients. After reading about the gender identity clinic in *Time* and *Newsweek* as a teen, Dana Beyer came to Hopkins to consider the surgery in the '70s. But when she had her first consultation, she blanched. "It was so highly sexualized, which was not at all my experience, certainly not the reason I was going to Hopkins to consider transition, that I just got up and left, I didn't want anything to do with it," said Beyer, who is now a board member of Gender Rights Maryland, in a 2014 interview for the *Johns Hopkins News-Letter*. "No one said this explicitly, but they certainly implied it, that the whole purpose of this was to get a vagina so you could be penetrated by a penis."

Instead, Meyer went to a private physician in Colorado decades later, in 2003.

———

Dr. Stanley Biber had been working as a small-town surgeon for fifteen years when he got the request that would change his life and transform his practice. It was 1969, and he had just gotten a visit from a woman he knew, a social worker who often accompanied young children born with cleft lips or palates to his office. For years, she'd been quietly impressed by his skill with the scalpel. Now she had a request of her own: Could he do her surgery?

"Well, of course," Biber replied. "What do you want done?"

"I'm a transsexual," the social worker replied.

Biber's mouth dropped. "What's that?" he asked.

He had never heard the term. The social worker explained that, while she had been born biologically male, she had been living as a woman and taking estrogen under the supervision of Dr. Harry Benjamin. She wanted Biber to complete her transformation by reshaping her penis and scrotum into a vagina and vulva.

As a general practitioner and the only surgeon in the town of ten thousand people, the forty-six-year-old Biber spent most of his time delivering babies, removing gallbladders, and performing appendectomies. He didn't have the first clue how to do such a surgery. But he agreed. "I wasn't very humble in those days," he would recall.

Trinidad, Colorado, is one of America's last true frontier towns. Three hours south of Denver and just a few miles north of the New Mexican border, it sits nestled in the long shadows of the Rocky Mountains' eastern foothills. Up on a high peak, a sign beams "Trinidad" in the style of the old Hollywood sign. The town was once a key trade stop on the Santa Fe Trail, attracting gold miners, missionaries, and other west-traveling settlers. In the 1800s it became a booming coal-mining town. As the mines closed, the economy dried up, and the population dwindled. But even down on its luck, Trinidad was a town of trailblazers, innovators, doers. And although Biber didn't know it, he was about to become one himself.

Trinidad was remote from any of the major clinics performing gender affirmation surgeries. To get an idea of what this surgery would entail, Biber called up Benjamin in New York, who referred him to the surgeons at Johns Hopkins. Hoopes, the head of the clinic, sent him hand-drawn notes and diagrams that detailed how to turn a man's genitalia into a woman's. The ink-and-pen sketches showed a similar surgery to the one Christine Jorgensen had back in 1952. Called the "penile scrotal-flap technique," it entailed the removal of the scrotum and the use of the penis to make a flap that would line the vagina. The technique, Biber recalled, was basic—crude, even. But he was able to use it as a guide.

And so, in the tiny forty-bed Mt. San Rafael Hospital originally run by Catholic nuns, Biber completed his first gender affirmation surgery. "It looked like hell, but it worked," he would later say. Most importantly, his patient was pleased with the result. Word soon spread that, against all odds, tiny Trinidad had a skilled surgeon who was willing to perform these surgeries.

By all accounts, Biber was an unlikely champion of gender affirmation surgery. Originally from Iowa, he had considered becoming a rabbi or a concert pianist. After a stint doing intelligence during World War II, he enrolled in medical school and became a surgeon, later leading a mobile army surgical hospital in the Korean War. When he returned home, he heard that the United Mine Workers of America was establishing a clinic for injured miners in Trinidad, Colorado. He decided to go west, thinking he'd work there for a year, maybe two. He never left. Seeing patients out of the First National Bank of Trinidad, a five-story sandstone building in the heart of the city's historic downtown, he became the city's only surgeon.

Biber adapted well to the rugged west. Standing at just five-foot-two, he came to work in muddy, tattered cowboy boots, blue jeans with a silver-buckled belt, and a cowboy hat over his balding pate. He'd taught himself to ride horses and drive pickup trucks, and lived on a sprawling ranch outside of town with his wife and children. He was also a lifelong body builder; legend had it that he'd once narrowly missed the cut for the US Olympic weightlifting team. He bragged that, behind enemy lines in the Korean War, he had once performed thirty-seven continuous abdominal surgeries before passing out.

Yet he had something that his patients badly wanted, something they didn't teach at Johns Hopkins: compassion. At the time, it was hard enough to find a surgeon who would treat you like a human being. In the 1960s, if you weren't lucky enough to get into Hopkins, you had to go to Mexico or Morocco, or risk a quack in a "chop shop." By contrast, Biber soon gained a reputation as a thoughtful, kind practitioner. "You have to see these people and know these people to have empathy with them," he said. And so, thanks to an open mind and a chance request, Biber swiftly rose to become the "dean of sex-change surgery."

Worried about the hospital's reaction, Biber initially hid the files of his first three genital surgeries in a safe in the administrator's office. But after giving lectures to religious leaders on gender dysphoria and gender affirmation surgery, he struck an uneasy balance. Though many residents didn't approve of his surgeries, the economic benefit to the town was clear. His surgeries brought $750,000 per year into the hospital coffers—not to mention the tourism that came with it. "It's a boon to business here," he told the *New York Times* in 1998. "They come with families, they stay in the hotels, they eat in the restaurants, they buy at the florists."

Soon patients flocked from nearly every part of the globe. His patients—"my transsexuals," he called them—included three brothers who "became" three sisters; a seventy-four-year-old-widower who had waited until his wife had died to complete his transition; an eighty-four-year-old who wanted to die as a woman; and Georgina Beyer, a member of New Zealand's parliament. "Movie stars, judges, mayors—everything," Biber said in 1998. "I had everything except a president of the United States."

Reporters, too, flocked to see it with their own eyes—and often reported back in sensationalist terms. "Someone who goes in male at 9 a.m. can be female by noon," the Associated Press wrote breathlessly in 1985. In 1993, TV talk-show host Geraldo Rivera came to Trinidad to observe one of Biber's surgeries on a twenty-two-year-old-patient. "I have never seen anything quite like it in my life, and I've seen quite a lot," Rivera said. Though he came with a camera crew, when it came to showing the actual operation, he balked. "I think it's a little too graphic for television," he said.

Biber wasn't shy about his accomplishments. "My work is the cat's whiskers," he told a reporter in 1984. "It's state of the art. You simply can't tell. Most of my patients can fool their gynecologists." His favorite boast was the woman he'd treated who was married to a gynecologist. Her husband, he claimed, didn't suspect a thing.*

His ascent was just in time. In 1979, Johns Hopkins abruptly closed its doors after a controversial report was released by its own doctors. The report, which faced criticism for its flawed methods and conclusions, found that of fifty patients surveyed, none were better off after their surgery than before. Its driving force was chief of psychiatry Paul McHugh, who concluded that, by conducting surgeries, "Hopkins was fundamentally cooperating with a mental illness." (He would later admit that he came to Hopkins with the explicit goal to end these surgeries.) That report sounded the death knell for university clinics.

For decades, Biber would be one of only a handful of unaffiliated doctors to perform these surgeries. But as university-run clinics closed around the country, a number of private physicians and surgical centers opened up to fill the void. By the 1970s there were gender affirmation surgeons in New York, Tucson, Jacksonville, and Chicago. For the first time, patients who would have been rejected by the university clinics could find surgeons willing to help them—provided they had the funds. Health insurance wouldn't cover gender affirmation surgery

---

* Most gynecologists I talked to say this is highly unlikely.

for decades, and Biber required his patients' fees—in the 1960s, about $3,225 to Biber and his team, another $3,000 to the hospital, for a total of around $52,000 in today's dollars—in cash.

By the 2000s, Biber had turned Trinidad into the "Sex-Change Capital of the World"—much to the chagrin of some of the town's more conservative residents. At one point, according to the *Los Angeles Times*, he boasted of doing 60 percent of the world's gender affirmation operations. In his thirty years, he estimated that he had performed five thousand male-to-female surgeries and eight hundred female-to-male surgeries. Getting gender affirmation surgery came to be known simply as "taking a trip to Trinidad."

As he entered his seventies, Biber had no thoughts of retiring. "As long as my hand's steady and my mind's clear, I'm going to continue to do transsexuals," he said in 1995. But by then there were more than a dozen well-known surgeons across the country, and Biber was seeing fewer and fewer patients. He performed his last gender affirmation surgery in 2003. Six months before he stopped doing surgeries, he recruited Marci Bowers from her successful OB/GYN practice in Seattle. Less than a year after Bowers joined the team, she was performing her first gender affirmation surgeries.

———

When Bowers flew to Colorado to meet Biber, he was already a legend. On May 25, 2000, the day she met him, he was on the cover of *USA Today* magazine. He'd invited her to spend some time with him and see his practice in the hopes that she would be his successor. "He planted the seed," she says. She'd been involved with the trans community in Seattle, doing hysterectomies and aftercare for some patients who had had vaginoplasties. But she wasn't quite convinced—there were several other job opportunities on the table.

Circumstances soon conspired to make Biber's offer seem more appealing. Back in Seattle, Bowers had recently gone through her own transition. While many of her OB/GYN patients and colleagues sent her flowers, some patients left because they weren't comfortable with her, she recalls. She applied for a job at a Washington state branch of Planned Parenthood, and was on the short list of candidates—until another physician interviewing her outed her as transgender, Bowers says. Soon after, she was told a polite "no, thank you." Next, she considered a position as the first female doctor at a small Christian OB/GYN group outside Seattle. Once again, she was outed during the interview process.

The Monday morning after she had returned from her first Harry Benjamin International Gender Dysphoria Association meeting (now the World Professional Association for Transgender Health, which Bowers became president of in 2022), she got to work to find a fax from the company withdrawing their employment offer. That's when she knew what she had to do. "It was clearly meant to be," she says now. She moved down to Colorado, where, as she puts it, "trucks outnumber people by about three to one."

In 2003, Bowers had never seen a gender affirmation surgery herself.* But as she watched Biber in the operating room, she started to feel like this was something within her surgical skill set. Little by little, she began taking over different parts of the procedure. Biber, for his part, knew immediately that he had found his successor. "Many have come to Trinidad to learn 'the surgery,'" he said in 2004, two years before he passed away. "But some have lacked the hands, some have lacked the confidence, and some have lacked the heart. Marci is the first to have all three."

Biber's surgeries were reliable, reproducible, and patients knew what they could expect. But Bowers soon saw opportunities to improve. In the 1960s, when Biber had started, physicians had very different attitudes toward female sexuality. Constructing a functioning, sensitive clitoris was more of an afterthought than a priority. For Bowers, the

---

* Although, of course, "I had some idea, that's for sure," she says.

clitoris was essential. "And so he had some serious deficiencies in the way he was doing it," she says. "But in a way he held the door open for me." In 2007, the *Denver Post* deemed her "Trinidad's transgender rock star."

Today, she says she does an operation that's about 80 percent different from what Biber did. Those innovations are mostly in the realm of appearance and pleasure. "I've taken it, I think, to a new level of anatomic reproducibility," she says.

But what does it mean to reproduce the female body? No matter how close Bowers comes to creating what she considers "normal" and "natural," she can never quite reach that goal—not due to a lack of technical skill, but because there is no such thing. The female body, as we have seen, is an idea, and an ideal. And Bowers has a new ideal in mind—one that's a lot more appealing than that of past anatomists. She shares the same vision as Marie Bonaparte, Aminata Soumare, and Roxanne Euber: one of wholeness, harmony, and completeness. She wants to take the orchestra of sexuality, switch out a few instruments, and write a new symphony—one that's slightly different but just as rich and complex.

———

When Roxanne Euber woke up from anesthesia, the first thing she felt was a strong urge to pee. "It didn't really hit me where I was for a quick second," she said. "It was like waking up from a really decadent nap." Then a nurse appeared to tell her she was done, it had all turned out well, and the memories came flooding back. "And then that's when the smile started," she said.

Roxanne was lying in her recovery suite on the fourth floor, where she'd just ordered some much-needed hospital room service: fruit and cottage cheese for her; salmon with lemon caper sauce and chocolate ice cream for Elle. Friends were just starting to check in. "Facebook, Instagram, you name it, it's all blowing up," she said. "People are just freaking out, it's like, 'Oh my God, finally!'" When her food came, she emptied a packet of salt into her cottage cheese. She chewed slowly, savoring the chunks of honeydew melon, cantaloupe, and pineapple.

She and Elle talked about everyday things: paying the bills, past boyfriends and girlfriends, how Roxanne used to blast Lady Gaga's "Born This Way" in her car to train her voice after transitioning.

All of a sudden Roxanne smiled and whispered conspiratorially:

"Guys, guess what? I have a vagina. Finally."

Elle rolled her eyes. "Great, now I'm gonna get text messages like 'Knock-knock.' 'Who's there?' 'I got a vagina.'"

"More importantly, that *thing* is gone forever," Roxanne said. "God, I hated that thing."

Roxanne knew this wouldn't be the end of her journey. New genitals, no matter how beautiful and well constructed, wouldn't solve all her problems. And yet genitals, despite (or perhaps because of) the outsized weight society places on them, do matter. "Everything is so focused on what's between your legs," she told me the first time I met her. "If I'd been born with webbed toes, nobody would care. But because I was born inside the body of the opposite sex, now it's everybody's friggin' business."

For now, at least, she was euphoric. "I finally fit my skin. My skin finally fits me," she said. "I'm deliriously happy."

AFTERWORD

On September 2, 2020, I got an email from someone named Bo Laurent. The name sounded vaguely familiar. Bo had seen an educational video I had made for *Scientific American* called "The Clitoris, Uncovered." In it, I point out anatomical features on a glowing, ghostly clitoris that hovers above my head and talk about some of the reasons it took so long for science to fully understand this remarkable organ. I start by mentioning a few of the reasons that knowing this anatomy is so important: it provides the foundation for surgeons who do gender affirmation surgery, as well as those who operate on women who have undergone genital cutting.

Bo was writing to encourage me to investigate another consequence of our anatomical ignorance about this part of the body: a form of genital cutting still being practiced right here in the United States, on children born with "unusual sex anatomy." It was similar to the form of clitorial amputation used to "cure" women of masturbation in the time of Marie Bonaparte, and to forms of genital cutting that women like Aminata and Aïssa went through. Bo knew this surgery intimately, because she was a survivor of it.* "Wikipedia will give you some sense of who I am," she added.

When I looked her up, I quickly realized why her name sounded familiar. Bo, who had previously published under the name Cheryl Chase, is the most well-known intersex activist in the world. I had read her work in a course called Science of Sex, Race, and Gender at the Massachusetts Institute of Technology a couple of years earlier. She was born in 1956 with atypical genitalia—in her case, a larger-than-usual clitoris. Doctors sent her home a boy. But one and a half years later, they opened up her abdomen and made a discovery: her gonads had both ovarian and testicular tissue. They jotted down in her medical files: "true hermaphrodite."

Then those doctors made a decision that would shape the course of Bo's life. First, they removed as much of her clitoral tissue as possible, and sewed up the skin around it. Then, they advised her parents to change her name, move to a new town, and raise her as a girl. Never speak of this time again. That was how Bo became Bonnie Sullivan.

For years after the procedure, Bo fell silent. When she found out as a young adult what had been done to her, her world fell apart. The surgery was bad: she would virtually never orgasm or feel genital pleasure ever again. But worse was the knowledge that the people in her life she'd trusted the most—her doctors, parents, family members—had lied to her. She would spend the next years in emotional turmoil. At twenty-one years old, she finally gained access to her medical records and found out the full story. But she still didn't have closure.

⁌

* Bo prefers the pronouns she/her, saying they fit her better, but she doesn't mind they/them, either. "I'm not a woman like most women," she says.

Eventually, Bo realized that while she couldn't change her anatomy, she could change her narrative. "The thing that they had done to me that was the worst was just to make me feel ashamed," she told me in October 2020, over Zoom. "And that was amenable to change."

She began reading Alfred Kinsey's sexuality studies, writing letters to surgeons who performed intersex surgeries, and attending "sex school"—the now-defunct Institute for Advanced Study of Human Sexuality—in San Francisco. Soon she realized that there were hundreds, maybe thousands, of others like her, each trapped in their own web of shame and lies. They had no way to reach out to one another. So she founded ISNA, the Intersex Society of North America, the first community of intersex people. Her goal was not only to provide support to intersex people but to change medical practice so no one would have to go through what she had.

It had taken nearly thirty years of medical activism, but in her first email to me, Bo told me about one of the first major triumphs of the modern intersex movement: Lurie, a children's hospital in Chicago, had become the first in the country to officially ban most intersex surgeries. For decades, performing "emergency" surgery on an infant's genitals was par for the course—not just expected but mandatory. Today, human rights groups are finally acknowledging the cruelty of inflicting radical, permanently scarring surgery on infants who cannot consent. That even infants have the right to pleasure, wholeness, and health.

Bo's story made something clear to me: anatomical knowledge doesn't stay within the academy. It radiates out along the nerve centers of society, through medicine, politics, and culture. It is one thing to say that shoddy science has led to incomplete textbooks and insufficient medical education. It is another to sit across from someone whose body, whose life, has been shaped by cultural notions of what a woman is and should be—and what medical interventions are appropriate to make her fit that mold. Bo helped me see what I could not: that the beliefs we share as a society about sex and gender harm all bodies. Culture, and medicine, shapes bodies.

———

In one of our Zoom conversations, Bo got out a five-centimeter ruler she had made that she'd dubbed the "Phall-O-Meter." Anything in the one-centimeter range was labeled SURGERY, overlaid with a sad-face emoji. On the left was the female symbol, and on the right, the male symbol. The point was to show how surgeons decide whether an intersex child will end up a boy or a girl. "It has to be bigger than 2.5 centimeters, stretched, to be a penis, and it has to be less than 0.9 centimeters to be a clitoris," said Bo. "Anything in between, just cut it off and call it a girl."

This is the basis by which doctors have assigned male and female bodies—not based on objective science, but based on deeply enmeshed ideas of what makes a man or a woman. In many cases, it still comes down to genitalia. We do not live in a world of "post-genital politics," to borrow a phrase from gender scholar Judith Butler. The kind of surgery that was performed on Bo also reflects the ancient assumption that the female body is the "default," the less-developed kind of body, the easier kind to manipulate. Almost all intersex surgeries turn babies into girls, by removing or burying clitoral tissue so there is no sign of a phallus. If the patient can have vaginal intercourse and, ideally, give birth, her surgery is considered "successful."

As Jocelyn Elders, a pediatric endocrinologist and the surgeon general under President Bill Clinton, once put it: "I can make a good female, but it's very hard to make a male." (She has since apologized.)

Today, debates still rage over who is "biologically a male" and "biologically a female"—and what rights people who don't fall into either category deserve when it comes to marriage, adoption, sports, bathrooms, and military service. Hormones, chromosomes, and high-tech "sex tests" have entered the mix. Often, one side appeals to "the science" to serve as the final word. But just like race, gender is a cultural framework that doesn't align smoothly with what we know about bodies and organs, or genetics and chromosomes.

From the moment you opened this book, you knew you weren't going to get an answer to the question of what makes a woman. We are still exploring the female body. We are still defining it, and then rewriting and expanding that definition as we go. The boundaries are fuzzier than ever. If science has anything to say about gender, it is that we are more alike than we are different; that gender is a spectrum, and that hormones, chromosomes, and genitals can arrange themselves in myriad permutations—"endless forms most beautiful," to borrow a phrase from Darwin. That if the female body is anything, it is a vehicle for change, and for blurring boundaries.

Just as Roxanne was not born with the privilege of having her genitalia match her identity, Bo was not born into one of the two boxes that society deemed acceptable or normal. Both felt firsthand the injustice of being judged and sorted, of having their potential curtailed, for the arbitrary fact of what was between their legs. Yet it was their bodies, and their lives spent in those bodies, that gave them the vantage point to see what others could not, to look from the outside at a system most of us take wholly for granted.

Our bodies can blind us. But they can also free us to see differently. They can help us bear witness to how a multitude of people, bodies, and perspectives have fallen through the cracks. Only by seeing connections instead of siloes, and sameness instead of difference, can we move the science of the female body forward and point the way to a truer, fuller understanding of all bodies.

ACKNOWLEDGMENTS

A book is a bit like the clitoris. The part you can see and touch—the book you hold in your hands—is only about 10 percent of the real thing. Beneath it swells a tingling matrix of friends, family members, colleagues, and generous strangers who lent their minds and hearts to this project. They are the nerves, the flesh, and the blood vessels; without them, this book would be an empty shell.

I want to first thank the remarkable scientists who allowed me into their work and lives. They include urologist Helen O'Connell, gynecologist and surgeon Ghada Hatem, biologist Patty Brennan, vaginal microbiome researcher Caroline Mitchell, ovarian biologists Dori Woods and Jonathan Tilly, bioengineer Linda Griffith, and gender affirmation surgeon Marci Bowers. Not only did they share their expertise and research with me, but they gamely let me turn the microscope on them, engaging with questions far beyond their scope of research and reflecting on the less-than-scientific forces that shaped their work and worldview. I am indebted to their openness and generosity of spirit, and I hope they will find that I was fair and true to them in my writing.

I also want to thank those who generously shared their personal journeys and their experiences with the medical system, as difficult as it was to do so. Bo Laurent, Roxanne Euber, Aminata Soumare, Aïssa Edon, "Alma," Victoria Field, Maisha Johnson, Jaipreet Virdi, and Cori Smith spent days and weeks patiently helping me understand journeys they had spent a lifetime navigating. They may not have intended to become explorers of the female body, but circumstances pushed them into anatomical quests of a different kind. Faced with medical ignorance, disinterest, and even hostility, they forged their own paths forward into understanding and embodied knowledge.

Archivists and librarians, my personal heroes, made it possible for me to piece together the lives of historical figures like Marie Bonaparte and Miriam Menkin. The staff of the Harvard Countway Library's Center for the History of Medicine—in particular Stephanie Krauss, Jessica Murphy, and Scott Podolsky—patiently guided me through the Menkin papers. Elizabeth Lockett, collection manager at the National Museum of Health & Medicine, walked me through Menkin's scientific legacy. Library of Congress archivist Margaret McAleer shared my enthusiasm for the Marie Bonaparte files and helped me navigate a trove of newly released materials, from sketches to journal entries. Librarian Jennifer Greenleaf, MIT's resident expert on women and gender studies, ensured that I had crucial university library access and personally dug up several hard-to-find documents.

I am deeply grateful to the researchers who reviewed the manuscript and pushed me to imagine new possibilities for this book. Biological anthropologist Heather Shattuck-Heidorn, of the University of Southern Maine, showed me the ways in which gender theory can do more than critique scientific knowledge: it can guide, shape, and create it. Reproductive anthropologist Kate Clancy and current and former members of her lab at the University of Illinois Champaign-Urbana—Valerie Sgheiza, Merri Wilson, Katie Lee, and Emma Verstraete—brought an immense breadth of knowledge to the project. They helped me identify some of my own blind spots and omissions, and begin the work of filling them in. Women's health nurse and HIV researcher

Jessica Wells of Emory University brought a critical eye to matters of gender, health, and race. Scholars Elizabeth Reis, Banu Subramanium, Deirdre Cooper Owens, Susan Stryker, Sarah Richardson, and Randi Epstein thoughtfully reviewed individual chapters and sections.

I am grateful to the Knight Science Journalism Program for allowing me to immerse myself in classes on gender theory and reproductive biology, and put the two into conversation. Program staff Deborah Blum, Ashley Smart, and Bettina Urcuioli helped shape and guide my project in 2018. Journalism fellows Pakinam Amer, Magnus Bjerg, Talia Bronstein, Jason Dearen, Lisa De Bode, Tim De Chant, Jeffery DelViscio, Elana Gordon, and Amina Khan shared their ideas, wisdom, and many plates of nachos. Jason became my book mentor, spirit guide, and lifelong friend. Jeff made my dreams of creating a 3D-clitoris video come true. Lisa reminded me why this book was urgent, and necessary.

I am also eternally beholden to the Alfred P. Sloan Foundation, which made it possible for me to do additional research and fact-checking, thus ensuring that this book was as accurate as possible; any errors are entirely my own. I am grateful to the MacDowell Fellowship Program, which granted me four magical weeks in the woods of New Hampshire and invaluable connections with other artists and writers.

I am indebted to the marvel of a fact-checker that is Kelsey Kudak. Kelsey not only caught my many errors and omissions, but dug up reporting notes and archival material I couldn't have dreamed of. I am privileged to have worked alongside her on this book.

Although I am running out of synonyms for grateful, I am truly grateful for my editor at W. W. Norton & Company, Melanie Tortoroli, who gave me a steady stream of support and fielded my many questions about the publishing industry. She enfolded my manuscript like a favorite bra: supportive, uplifting, and firm where necessary. Assistant editor Mo Crist was an invaluable reader, challenging my assumptions and asking incisive questions that helped lead me down fruitful rabbit holes. My literary agent, Danielle Svetcov, of the Levine Greenberg Rostan Literary Agency, guided me through the publishing world and helped keep me focused on the bigger picture beyond the weeds. Thanks to her, I was able to glimpse a vision of what this book could be. Laura Helmuth, my first mentor and editor at *Slate*, valued me as a science reporter and made me feel like I had something to say. My editor at the *New York Times*, Alan Burdick, believed in my stories and my voice, whether I was writing about sperm science or chlamydia koalas.

My eternal gratitude goes out to my beloved friends and family, many of whom I unfairly pressed into editorial service. Liz Vogt, Jacqueline Mansky, Lorraine Boissoneault, and Jacqueline Berkman all read my rawest drafts and made them substantially better. Lorraine and Catherine Bennett provided critical translation and reporting help. My sister, Aubrey Gross; my sister-in-law, Nur Ibrahim; my father, Mark Gross; and my stepmother, Weimin Sun, leant me support while also allowing me to regale them with every vagina fact I learned. My brother, journalist Daniel A. Gross, was an invaluable source of inspiration. His reporting, storytelling, and keen sense of justice have long been a lodestar for me. My mother, Désirée Lie, MD, supported me unrelentingly. She brought all her skills as a mother, teacher, writer, editor, and doctor, leaping over every gulf of generational and emotional understanding to be—quite literally—on the same page as me.

To everyone involved in the making of this book: Thank you for bringing your wild imaginations, your deep knowledge stores, and your boundless compassion. This book, and I, am incalculably better for it.

ENDNOTES

Introduction: Named, Claimed, and Shamed

x **"Boric acid is a dangerous poison":** Anthony P. Restuccio et al., "Fatal ingestion of boric acid in an adult." *The American Journal of Emergency Medicine* 10, no. 6 (1992): 545–547. DOI: 10.1016/0735-6757(92)90180-6.

xi **a federal mandate required researchers:** "History of Women's Participation in Clinical Research," Office of Research on Women's Health, National Institutes of Health, orwh.od.nih.gov/toolkit/recruitment/history. (To be fair, the NIH encouraged this policy since 1986, but it was not required by law.)

xii **"What part of NIH is interested in lady parts?":** Diana Bianchi, interview by the author, September 14, 2020.

xiii **the term was officially retired:** Rachel E. Gross, "Taking the 'Shame Part' out of Female Anatomy," *New York Times*, September 21, 2021, www.nytimes.com/2021/09/21/science/pudendum-women-anatomy.html.

xiv **Surveys find:** Kate Watson, "Way Too Many Women Don't Know Where Their Vaginas Are," *Vice*, September 8, 2016; Jade Bremmer, "Quarter of American Women Don't Know Where Their Vagina Is," *Independent*, November 10, 2020.

xvi **biology has proven this not to be the case:** Claire Answorth, "Sex Redefined: The Idea of 2 Sexes Is Overly Simplistic," *Nature* 518 (2015): 288–291.

Chapter 1: Desire

Most of the quotes and personal details about Marie's life are taken from Marie Bonaparte's magnum opus, *Female Sexuality* (1953); Celia Bertin's biography *Marie Bonaparte: A Life* (1992); and Marie Bonaparte's extensive papers at the Library of Congress,
the bulk of which were opened January 1, 2020. French translation was provided by
journalist Lorraine Boissoneault.

1 **article advocating for it in a medical journal:** A. Narjani, "Considerations sur les causes anatomiques de frigidité chez la femme," *Bruxelles-Medical*
   27 (1924): 768–778.

1 **"the suspensory ligament":** Marie Bonaparte, *Female Sexuality* (New York:
   International Universities Press, Inc., 1951), 151.

2 **"I hate it as much as you do":** Celia Bertin, *Marie Bonaparte* (Paris: Perrin,
   1992), 94.

2 **"My libido is all in my head":** Bertin, *Marie,* 184.

2 **not exactly a routine occurrence:** Survey findings of how many women
   orgasm from intercourse alone vary, but most find it hovers between 20 and 30
   percent.

   In a 2017 nationally representative survey of more than 1,000 women by
   Debby Herbernick et al., 18 percent said intercourse alone was sufficient for
   orgasm, while an additional 36 percent said clitoral stimulation enhanced
   their orgasm. (However, more than half of women in the former category
   experienced orgasm "infrequently.") Meanwhile, in an online survey of
   1,500 women, Kim Wallen et al. found that, on average, women reported
   orgasm during "unassisted" intercourse 21 to 30 percent of the time, whereas
   with "assisted" intercourse (meaning, with clitoral stimulation) it was 51 to
   60 percent.

   In her now-classic 1976 survey of more than 3,000 American
   women, sex researcher Shere Hite also found that fewer than 30 percent of
   women experienced orgasm from intercourse alone. Hite concluded that one
   reason for this discrepancy was the fact that women were so often forced to fit
   their sexual experience into men's expectations and scripts. "For a woman to
   have orgasm during intercourse, from intercourse, is simply not the majority
   experience," she wrote. "Even the question being asked is wrong. The question should not be: Why aren't women having orgasms from intercourse? But
   rather: Why have we insisted women should orgasm from intercourse?"

Complicating this data yet further is the fact that each survey asks the question differently (for instance, "what is your most reliable route to orgasm" versus what do you need "in order to orgasm during intercourse"?), which can skew responses. Imprecise language has also been shown to muddy the waters: in particular, surveys that ask whether women orgasm "during intercourse" often do not specify whether "intercourse" includes clitoral stimulation. Finally, cultural pressure may lead some women to report that they orgasm from penetration alone when they do not—the statistical version of faking it.

Debby Herbenick et al., "Women's Experiences with Genital Touching, Sexual Pleasure, and Orgasm: Results from a U.S. Probability Sample of Women Ages 18 to 94," *Journal of Sex and Marital Therapy* 44, no. 2 (August 2017): 201–212.

Shere Hite, *The Hite Report: A Nationwide Study of Female Sexuality* (New York: Macmillan Publishing Co., 1976), 181–189.

Talia Shirazi et al., "Women's Experience of Orgasm During Intercourse: Question Semantics Affect Women's Reports and Men's Estimates of Orgasm Occurrence," *Archives of Sexual Behavior* 47, no. 3 (2018): 605–613.

2 **"la Grande Nation"**: Marie-Monique Huss, "Pronatalism in the Inter-War Period in France," *Journal of Contemporary History* 25, no. 1 (1990): 39.

2 **"normal, vaginal, maternal"**: Marie Bonaparte, *Female Sexuality* (New York: International Universities Press, Inc., 1951), 1.

3 **"prouder of her masturbation"**: Bertin, *Marie*, 161.

3 **twenty-two minutes**: Bertin, *Marie,* 170.

4 **"the penis and orgastic normality"**: Bertin, *Marie,* 157.

5 **left her wedding ring**: Bertin, *Marie,* 157.

6 **Well-adapted women**: Sigmund Freud, "Female Sexuality," in *The Standard Edition of the Complete Psychological Works of Sigmund Freud*, vol. 21 (London: Hogarth Press and the Institute of Psycho-Analysis, 1953), 221–244.

6 **"phallic woman"**: Nellie L. Thompson, "Marie Bonaparte's Theory of Female Sexuality: Fantasy and Biology," *American Imago* 60, no. 3 (2003): 349.

7 **remove an ovarian cyst**: Bertin, *Marie*, 146.

7 **the vagina reigned supreme**: Thomas Laqueur, *Making Sex: Body and Gender from the Greeks to Freud* (Cambridge, MA: Harvard University Press, 1990), 4.

7 **some anatomical truth to homology**: Saladin, *Human Anatomy*, 1124–1126.

7 **At six weeks after conception**: Kenneth Saladin, *Human Anatomy*, 5th ed. (New York: McGraw-Hill Education, 2017), 727–729.

8 **"by default"**: Saladin, *Human Anatomy*, 729.

8   **a suite of active factors:** Jerome F. Strauss, Robert L. Barbieri, and Samuel
     S. C. Yen, *Yen & Jaffe's Reproductive Endocrinology: Physiology, Pathophysiology,
     and Clinical Management,* 8th ed. (Philadelphia: Elsevier, 2019), 170.

8   **clitoris grows inward:** Vincent Di Marino and Hubert Lepidi, *Anatomic
     Study of the Clitoris and the Bulbo-Clitoral Organ* (Springer International Publish-
     ing, 2014).

9   **"This small formation is called the nymph":** Vincent Di Marino and
     Hubert Lepidi, *Anatomic Study of the Clitoris and the Bulbo-Clitoral Organ*
     (Springer International Publishing, 2014), 3.

9   **remnants of each sex remain in the other:** Gary Cunningham et al., *Wil-
     liams Obstetrics,* 25th ed. (New York: McGraw-Hill Education, 2018), 33–38.

9   **structure called the appendix vesiculosa:** Johannes Sobotta, W. Hersey
     Thomas, and James Playfair McMurrich, *Atlas and Text-Book of Human Anatomy*
     (Philadelphia: W. B. Saunders Company, 1909), 148.

10   **temporary wooden theater:** Jonathan Sawday, *The Body Emblazoned: Dissec-
     tion and the Human Body in Renaissance Culture* (London, New York: Routledge,
     1995), 66–72.

11   **He continued to rely on:** Katharine Park, *Secrets of Women: Gender, Gen-
     eration, and the Origins of Human Dissection* (New York: Zone Books, 2010),
     218–219.

13   **an extensive organ richly supplied:** Thomas P. Lowry, ed., *The Classic
     Clitoris: Historic Contributions to Scientific Sexuality.* (United States: Nelson-Hall,
     1978): 20–24.

14   **"It's a sin!":** Thompson, "Marie Bonaparte's Theory," 347.

14   **"in the interest of the reproductive health":** Alison M. Moore, "Vic-
     torian Medicine Was Not Responsible for Repressing the Clitoris: Rethink-
     ing Homology in the Long History of Women's Genital Anatomy," *Signs* 44,
     no. 1 (2018): 68. For a full discussion on the (re)construction of femininity in
     interwar France, see: Marie-Louise, Roberts, *Civilization without Sexes: Recon-
     structing Gender in Postwar France, 1917–1927* (Chicago: University of Chicago
     Press, 1994).

14   **"women who allowed themselves to be led astray":** Moore, "Victorian
     Medicine," 63.

15   **applying blistering doses of carbolic acid:** J. H. Kellogg, *Plain Facts for Old
     and Young: Embracing the Natural History and Hygiene of Organic Life* (Burlington,
     IA: I. F. Segner, 1886), 296.

15   **"afflicted with compulsive masturbation":** Bonaparte, *Female Sexuality,* 157.

15 **clitoral procedures were also advertised:** Sarah B. Rodriguez, *Female Circumcision and Clitoridectomy in the United States: A History of a Medical Treatment* (Rochester, NY: Boydell & Brewer, 2014), 12.

16 **she was peeling an apple:** Adam Phillips, *Becoming Freud: The Making of a Psychoanalyst* (New Haven: Yale University Press, 2014), 73.

17 **"in the fifty-three years of our marriage":** Peter Gay, *Freud: A Life for Our Time* (New York: W. W. Norton, 2006), 60.

18 **an infantile organ:** Sigmund Freud and Elisabeth Young-Bruehl, *Freud on Women: A Reader* (New York: W. W. Norton, 1992), 347.

18 **"The great Darwin":** Lucille B. Ritvo, *Darwin's Influence on Freud: A Tale of Two Sciences* (New Haven: Yale University Press, 1990), 3.

18 **"Anatomy has recognized":** Freud and Young-Bruehl, *Freud on Women*, 158.

19 **"Every great thinker":** Richard Gilman, "The FemLib Case Against Freud," *New York Times,* January 31, 1973. (As an aside, this article contains the second recorded mention of the word "clitoris" in the Gray Lady.)

19 **"dark continent for psychology":** Sigmund Freud, "The Question of Lay Analysis," in *The Standard Edition of the Complete Psychological Works of Sigmund Freud,* vol. 20 (London: The Hogarth Press, 1959), 177–258.

20 **"erotic life":** Sigmund Freud, *Three Essays on the Theory of Sexuality.* (London: Imago Pub. Co, 1905), 151.

20 **"an impenetrable obscurity":** Freud and Young-Bruehl, *Freud on Women*, 98.

20 **"the great question":** Ernest Jones, *The Life and Work of Sigmund Freud,* vol. 2 (London: Hogarth Press, 1955), 421.

21 **"Nature and life gave me":** Marie Bonaparte Papers, Box 17, Folder 13.

21 **she had little satisfaction:** Bertin, *Marie,* 120–124.

21 **queried 243 women:** A. E. Narjani, "Considerations sur les causes anatomiques de frigidité chez la femme," *Bruxelles-Médical* 27 (1924): 768–778.

22 **a simple diagram of a vagina:** Marie Bonaparte Papers, Library of Congress, Box 22, Folder 8.

22 **measured herself:** Bonaparte, *Female Sexuality.*

22 **too distant to ensure orgasm:** Marie Bonaparte, "Les Deux Frigidités de la femme," *Bulletin de la Société de Sexologie* 1 (1932): 161–170.

22 **clitoral-vaginal distance does indeed play a role:** Mary Roach, *Bonk: The Curious Coupling of Science and Sex.* (New York: W. W. Norton, 2008), 68.

22 **women's tendency to orgasm:** Kim Wallen and Elisabeth A. Lloyd, "Female Sexual Arousal: Genital Anatomy and Orgasm in Intercourse," *Hormones and Behavior* (May 2011): 780–792.

23 **the sound of blackbirds:** Marie Bonaparte Papers, Box 17, Folder 12.

23 **"I believe Halban did great work!":** Marie Bonaparte Papers, Box 17, Folder 12.

23 **a small infection in her surgical wound:** Marie Bonaparte Papers, Box 17, Folder 12.

23 **a dreamlike confusion:** Marie Bonaparte Papers, Box 17, Folder 12.

24 **"*volle Befriedigung*":** Marie Bonaparte Papers, Box 17, Folder 12.

24 **"pre-analytical and erroneous":** Bonaparte, *Female Sexuality*, 150.

26 **"harmonious collaboration":** Bonaparte, *Female Sexuality*, 170.

26 **"splendid old thing":** Thompson, "Marie Bonaparte's Theory," 343.

26 **Scholars have lamented:** Daniel J. Fairbanks, "Mendel and Darwin: Untangling a Persistent Enigma," *Heredity,* December 17, 2009.

Chapter 2: Wholeness

Journalist Catherine Bennett translated interviews with clitoral surgeons in Paris and French-speaking women at La Maison de Femmes.

29 **Helen O'Connell's voice is tinged with awe:** Helen O'Connell, interview with author, February 28, 2020.

29 **tightly packed in the glans clitoris:** Cheryl Shih, Christopher Cold, and Claire Yang, "Cutaneous Corpuscular Receptors of the Human Glans Clitoris: Descriptive Characteristics and Comparison with the Glans Penis," *Journal of Sexual Medicine* 10, no. 7 (July 2013): 1783–1789.

30 **O'Connell's doctoral thesis:** Helen Elizabeth O'Connell, *Review of the Anatomy of the Clitoris,* Thesis for the Degree of Doctor of Medicine, University of Melbourne, April 2004.

30 **"The original anatomists weren't interested":** Sharon Mascall, "Time for Rethink on the Clitoris," *BBC,* June 11, 2006.

31 **"failure":** Raymond Jack Last, *Anatomy: regional and applied* (Edinburgh: Churchill Livingstone, 1985), 354–355.

31 **a long, winged body:** Suzann Gage, Carol Downer, and Rebecca Chalker, "The Clitoris: A Feminist Perspective," in *A New View of a Woman's Body* (West Hollywood, CA: Feminist Health Press, 1981), 33–57.

32 **"For a surgeon":** Rachel Nowak and Susan Williamson, "The Truth About Women," *New Scientist,* August 1, 1998.

32   **"a bit like a pair of shorts":** Megan Rees, interview by the author, February 26, 2020.

32   **their clitorises had shrunken somewhat with age:** Helen O'Connell, "Get Cliterate," TEDxMacRobHS, September 2020, www.ted.com/talks/professor_helen_o_connell_get_cliterate.

33   **hours to dissect just millimeters of flesh:** Rob Plenter, interview by the author, February 29, 2020.

34   **a set of magnetic resonance imaging (MRI) images:** Helen O'Connell and John O. DeLancey, "Clitoral Anatomy in Nulliparous, Healthy, Premenopausal Volunteers Using Unenhanced Magnetic Resonance Imaging," *The Journal of Urology* 173, no. 6 (June 2005), 2060–2063.

34   **many medical textbooks still referred:** Lisa Jean Moore and Adele E. Clarke, "Clitoral Conventions and Transgressions: Graphic Representations in Anatomy Texts, c1900–1991," *Feminist Studies* 21, no. 2 (Summer 1995): 255–301.

34   **ten times the size:** Helen O'Connell, Kalavampara V. Sanjeevan, and John M. Hutson, "Anatomy of the Clitoris," *The Journal of Urology* 174, no. 4 Pt 1 (October 2005), 1189–1195.

35   **"clitoris guru":** Fyfe, Melissa, "Get Cliterate: How a Melbourne Doctor Is Redefining Female Sexuality," *Sydney Morning Herald*, December 8, 2018.

35   **"Penis envy may be a thing of the past":** Rachel Nowak and Susan Williamson, "New Study of the Clitoris Reveals Truths Missed by Anatomy Textbooks," *New Scientist*, July 31, 1998.

35   **restored "the clitoris":** "Anatomy of a Revolution," *Sydney Morning Herald*, September 8, 2005.

35   **"in all its glory":** Norman Eizenberg, interview by the author, February 27, 2020.

35   **"passive female sex organ":** Thomas P. Lowry, ed., *The Classic Clitoris: Historic Contributions to Scientific Sexuality* (United States: Nelson-Hall, 1978), 22.

36   **She was sitting in a biology class:** Aminata Soumare, interview by the author, November 22, 2019. (Translation by Catherine Bennett.)

37   **scholars have convincingly argued:** Brian Earp et al., "The Need for a Unified Ethical Stance on Child Genital Cutting," *Nursing Ethics* (March 2021); Nancy Ehrenreich and Mark Barr, "Intersex Surgery, Female Genital Cutting, and the Selective Condemnation of 'Cultural Practices,'" *Harvard Civil Rights-Civil Liberties Law Review* 71 (2005); Brian D. Earp and Sara Johnsdotter,

"Current Critiques of the WHO Policy on Female Genital Mutilation," *International Journal of Impotence Research* 33 (2021): 196–209; Cheryl Chase, " 'Cultural Practice' or 'Reconstructive Surgery'? U.S. Genital Cutting, the Intersex Movement, and Medical Double Standards," in *Genital Cutting and Transnational Sisterhood*, eds. Stanlie M. James and Claire C. Robertson (Urbana, IL: University of Illinois Press, 2002), 145–146.

38 **8 in 10 girls in Mali:** UNICEF, "Mali: Statistical Profile on Female Genital Mutilation," January 2019, available at: www.ecoi.net/en/document/2025689.html.

38 **consent at age fifteen:** US Department of State, Bureau of Consular Affairs, Country Profile on Republic of Mali, travel.state.gov/content/travel/en/ international-travel/International-Travel-Country-Information-Pages/Mali .html/.

39 **"This 'operation' doesn't touch":** Sokhna Fall Ba, interview by the author, November 20, 2019. (Translation by Catherine Bennett.)

39 **"Our ancestors were no scientists":** Quoted in Fran P. Hosken, *The Hosken Report: Genital and Sexual Mutilation of Females* (Lexington, MA: Women's International Network News, 1979), 192–202.

39 **"You pray that it's over":** Corinne, interview by the author, November 18, 2019. (Translation by Catherine Bennett.)

40 **La Maison de Femmes is a pop of color:** Reporting details gathered in person at La Maison de Femmes on November 18, 2019.

40 **long been a silent problem:** Kim Willsher, "'No More Shame': The French Women Breaking the Law to Highlight Femicide," *The Guardian,* March 23, 2021.

42 **Hatem listened:** Ghada Hatem, interview by the author, November 18, 2019. (Translation by Catherine Bennett.)

43 **Hatem herself had emigrated from Lebanon:** Catherine Robin, "Ghada Hatem-Gantzer, la Dr House des femmes," *Elle,* www.elle.fr/Societe/Interviews/ Ghada-Hatem-Gantzer-la-Dr-House-des-femmes-2867670.

46 **"women feel the fact":** Pierre Foldès, Béatrice Cuzin, and Armelle Andro, "Reconstructive Surgery after Female Genital Mutilation: A Prospective Cohort Study," *Lancet* 380, no. 9837 (July 14, 2012): 134–141.

46 **"the everlasting gonad urge":** R. L. Dickinson, *Human Sex Anatomy: a Topographical Hand Atlas*, 2nd edition (Baltimore: Williams & Wilkins Company, 1949), VII.

46 **a British paleo-botanist named Marie Stopes:** Peter M. Cryle and Elizabeth Stephen, *Normality: A Critical Genealogy* (Chicago: Chicago University Press, 2017), 286–287.

47 **primarily utilizing the missionary position:** R. L. Dickinson, *A Thousand Marriages; A Medical Study of Sex Adjustment* (Baltimore: The Williams & Wilkins Company, 1932), page 66.

47 **the clitoris may be more densely innervated than the penis:** Shih, Cold, and Yang, "Cutaneous Corpuscular Receptors."

47 **Not all of these observations:** Roy Porter and Mikulas Teich, eds., *Sexual Knowledge, Sexual Science* (United Kingdom: Cambridge University Press, 1994), 311.

48 **white alabaster:** Rose Holz, "The 1939 Dickinson-Belskie Birth Series Sculptures: The Rise of Modern Visions of Pregnancy, the Roots of Modern Pro-Life Imagery, and Dr. Dickinson's Religious Case for Abortion," *Journal of Social History* 51, no. 4 (Summer 2018): 980–1022.

49 **within five years, 6.5 million women were on it:** Jonathan Eig, *The Birth of the Pill: How Four Crusaders Reinvented Sex and Launched a Revolution* (United Kingdom: W. W. Norton, 2014), 313.

52 **There was only one kind of orgasm:** William Masters and Virginia Johnson, *Human Sexual Response* (New York: HarperCollins Publishers, 1981), 66–67.

53 **Aïssa was six:** Mundasad Smitha, "The Midwife Who Is Trying to Save Women from FGM," *BBC News*, November 24, 2015, www.bbc.com/news/health-34809550.

54 **suggested by a man:** Melissa Fyfe, "Get Cliterate: How a Melbourne Doctor Is Redefining Female Sexuality," *Sydney Morning Herald*, December 8, 2018.

54 **He found himself perturbed:** Ernst Gräfenberg, "The Role of the Urethra in Female Orgasm," *International Journal of Sexology* 3 (1950): 145–148.

55 **"Missionary position just doesn't do it":** Wendy Zuckerberg, "The G-Spot," *Science Vs*, podcast, Gimlet Media, September 1, 2016.

56 **an anatomical UFO:** Rebecca Chalker, *The Clitoral Truth*, 2nd ed. (New York: Seven Stories Press, 2018), 48.

56 **"a deep, deep structure":** Melissa Healy, "Doctor Says He's Found the Actual G-Spot," *Sydney Morning Herald*, April 26, 2012.

56 **"With the stiffening of its walls":** Lowry, *Classic Clitoris*, 47.

56 **"the shape of a tremendously distended leech":** Lowry, *Classic Clitoris*, 25.

57 **the G-spot seemed to press:** Nathan Hoag, Janet R. Keast, and Helen E. O'Connell, "The 'G-Spot' Is Not a Structure Evident on Macroscopic Anatomic Dissection of the Vaginal Wall," *Journal of Sexual Medicine* 12 (December 2017): 1524–1532.

58 **"it doesn't work like that"**: Helen O'Connell, interview by the author, February 28, 2020.

59 **"That it doesn't just rely"**: Beverly Whipple, interview by the author, May 7, 2021.

Chapter 3: Resilience

61 **"You could die in that forest"**: Patty Brennan, interview by the author, May 7, 2021.

62 **Ninety-seven percent of all bird species**: Carl Zimmer, "In Ducks, War of the Sexes Plays Out in the Evolution of Genitalia," *New York Times*, May 1, 2007.

63 **some of the most wildly varying organs**: Patricia L. R. Brennan and Richard O. Prum, "Mechanisms and Evidence of Genital Coevolution: The Roles of Natural Selection, Mate Choice, and Sexual Conflict," *Cold Spring Harbor Perspectives in Biology* 7, no. 7 (July 2015): a017749.

63 **penises that can taste**: Emily Willingham, *Phallacy: Life Lessons from the Animal Penis* (New York: Avery, 2020), 80.

64 **"Actually the vagina"**: Anne Koedt, *Myth of the Vaginal Orgasm* (Boston: New England Free Press, 1970).

65 **"These are two extremely closely involved structures"**: Emily Willingham, interview by the author, February 17, 2021.

66 **focused almost entirely on the male**: Willingham, *Phallacy*, 129.

67 **"The secret was in the dissection"**: Patricia L. R. Brennan and Tim R. Birkhead, "Elaborate Vaginas and Long Phalli: Post-Copulatory Sexual Selection in Birds," *Biologist*, Institute of Biology 56, no. 1 (February 2009), 35.

68 **the longest known bird phallus**: Kevin McCracken et al., "Sexual Selection: Are Ducks Impressed by Drakes' Display?" *Nature* 413, no. 6852 (September 2001): 128.

68 **"It was fitting"**: Kevin McCracken, interview by the author, February 23, 2021.

68 **analyzed the vaginas of sixteen species**: Patricia L. R. Brennan et al., "Coevolution of Male and Female Genital Morphology in Waterfowl," *PLoS One* (May 2007), journals.plos.org/plosone/article?id=10.1371/journal.pone .0000418.

68 **"I got to thinking"**: Kevin McCracken, email to the author, March 4, 2021.

69 **"When you dissected one of the birds"**: Zimmer, "In Ducks."

69 **a struggle for reproductive control**: Richard Prum, "Duck Sex and the Patriarchy," *The New Yorker*, 2017; Patricia Brennan and Richard Prum, "The

Limits of Sexual Conflict in the Narrow Sense: New Insights from Waterfowl Biology," Philosophical Transactions of the Royal Society B, 2012.

70 **"very acute of Mr. Ruskin":** "Darwin in Letters, 1879," Darwin Correspondence Project, www.darwinproject.ac.uk/letters/darwins-life-letters/iable-letters-1879-tracing-roots/.

70 **"these parts are more brightly coloured":** Charles Darwin, "Sexual Selection in Relation to Monkeys," Nature 15 (1876): 18.

70 **"of a brilliant carmine red":** Charles Darwin, The Descent of Man: And Selection in Relation to Sex (London: J. Murray, 1871), 539.

72 **"Darwinians had to be like Caesar's wife":** Evelleen Richards, interview by the author, January 3, 2021.

72 **he cited reports by European colonists:** Evelleen Richards, Darwin and the Making of Sexual Selection (Chicago: The University of Chicago Press, 2017), 396.

72 **backsides of KhoiKhoi women:** Sadiah Qureshi, "Displaying Sara Baartman, the 'Hottentot Venus," History of Science 42, no. 2 (June 2004): 233–257; Sabrina Strings, Fearing the Black Body: The Racial Origins of Fat Phobia (New York: New York University Press: 2019).

72 **"the very girdle or protuberance":** Gowan Dawson, Darwin, Literature and Victorian Respectability. (Kiribati: Cambridge University Press, 2007), 38.

72 **had long attracted the prurient attention:** Anne Fausto-Sterling, "Gender, Race, and Nation: The Comparative Anatomy of 'Hottentot' Women in Europe, 1815–1817," in Deviant Bodies, eds. J. Terry and J. Urla (Bloomington: Indiana University Press, 1995), 19–48.

72 **"Because they were not seen as human":** Banu Subramanian, interview by the author, May 3, 2021. For a deeper look at how the ghosts of racism and sexism continue to haunt evolutionary biology, see: Subramaniam, Banu, Ghost Stories for Darwin: The Science of Variation and the Politics of Diversity (Champaign: University of Illinois Press, 2014).

72 **hadn't set foot on the continent since:** Richards, Darwin and the Making, 396.

74 **he ran into a woman:** Keith Wilson, ed. A Companion to Thomas Hardy (United Kingdom, Wiley-Blackwell, 2009).

74 **each reinforced the other:** Jim Endersby, "Gentlemanly Generation: Pangenesis and Sexual Selection," in The Cambridge Companion to Darwin (Cambridge, UK: Cambridge University Press, 2003), 73.

74 **"Women have served all these centuries":** Virginia Woolf, A Room of One's Own (United Kingdom: Harcourt Brace Jovanovich, 1989), 35.

75 **"graduated from school or college":** Edward Hammond Clarke, Sex in Education: Or, A Fair Chance for the Girls. (United States: J. R. Osgood, 1873), 39.

75 **"the fundamental biological law":** Mary Roth Walsh, *"Doctors Wanted, No Women Need Apply": Sexual Barriers in the Medical Profession, 1835–1975* (New Haven and London: Yale University Press, 1977), 232.

76 **"It is impressive":** W. G. Eberhard, "Post-Copulatory Sexual Selection: Darwin's Omission and Its Consequences," *Proceedings of the National Academy of Sciences* 106, Suppl (2009): 10025–10032.

77 **"the field on which males compete":** T. R. Birkhead and A. P. Møller, *Sperm Competition and Sexual Selection* (San Diego: Academic Press, 1998), 96.

77 **"The idea that perhaps male genitalia were courtship devices":** William Eberhard, email to the author, January 5, 2021.

77 **she had tricks up her sleeve:** William G. Eberhard, *Female Control: Sexual Selection by Cryptic Female Choice* (Princeton: Princeton University Press, 1996,) 81.

79 **this multifaceted tube can only vary so much:** Brennan and Prum, "Mechanisms and Evidence."

79 **The authors analyzed 364 papers:** M. Ah-King, A. B. Barron, and M. E. Herberstein, "Genital Evolution: Why Are Females Still Understudied?" *PloS Biol* 12, no. 5 (2014): e1001851.

80 **Fox News anchor Shannon Bream:** Prum, "Duck Sex."

80 **"Duckpenisgate":** Asawin Suebsaeng, "The Latest Conservative Outrage Is About Duck Penis," *Mother Jones,* March 26, 2013.

80 **"Genitalia, dear readers":** Patricia Brennan, "Why I Study Duck Genitalia," *Slate,* April 2, 2013, slate.com/technology/2013/04/duck-penis-controversy-nsf-is-right-to-fund-basic-research-that-conservatives-misrepresent.html.

81 **symposium on penis diversity:** "The Morphological Diversity of Intromittent Organs," Society for Integrative and Comparative Biology Annual Meeting, January 4, 2016, sicb.burkclients.com/meetings/2016/symposia/intromittent.php/.

81 **a series of inner fleshy lids:** E. J. Slijper, *Whales: The Biology of the Cetaceans,* trans. A. J. Pomerans (New York, Basic Books: 1962), 356; Dara N. Orbach et al., "Patterns of Cetacean Vaginal Folds Yield Insights into Functionality," *PloS One* 12, no. 3 (2017): e0175037.

81 **Her inspiration for this technique:** Paula Pendergrass, interview by the author, September 19, 2021; Rose Eveleth, "The Failed Vagina Story," *The Last Word on Nothing,* July 28, 2016, www.lastwordonnothing.com/2016/07/28/the-failed-vagina-story/.

82 **"unprecedented" vaginal diversity:** Dara N. Orbach et al., "Genital Interactions During Simulated Copulation Among Marine Mammals," *Proceedings of the Royal Society B: Biological Sciences* 284, no. 1864 (2017): 20171265.

83   **these vaginas were so unique:** Dara Orbach, interview by the author, December 29, 2020.

84   **Although the evidence was slight:** Brian S. Mautz, et al., "Penis size influences male attractiveness," *Proceedings of the National Academy of Sciences* April 2013, 110 (17), 6925–6930.

84   **human males have relatively wide:** Alan F. Dixson, *Sexual Selection and the Origins of Human Mating Systems* (United Kingdom: OUP Oxford, 2009), 65.

85   **more than 300 percent:** James A. Ashton-Miller and John O. L. DeLancey, "On the Biomechanics of Vaginal Birth and Common Sequelae," *Annual Review of Biomedical Engineering* 11 (August 2009): 163–176.

85   **13.75 inches in circumference:** "Newborn Measurements," Stanford Children's Health, www.stanfordchildrens.org/en/topic/default?id= measurements-90-P02673.

85   **six to twelve weeks:** Miranda A. Farage and Howard I. Maibach, eds., *The Vulva: Physiology and Clinical Management* (Boca Raton, FL: CRC Press, 2017), 18.

85   **"It's pretty straightforward":** Holly Dunsworth, "Why Is the Human Vagina So Big?" The Evolution Institute, December 3, 2015; Holly Dunsworth, "Why Is No One Interested in Vagina Size?" *New York Magazine,* December 16, 2015.

85   **The clitoris may also change shape:** Cesare Battaglia et al., "Morphometric and Vascular Modifications of the Clitoris During Pregnancy: A Longitudinal, Pilot Study," *Archives of Sexual Behavior* 47, no. 5 (2018): 1497–1505. DOI: 10.1007/s10508.017.1046-x.

86   **other potential explanations:** Holly Dunsworth, "Expanding the Evolutionary Explanations for Sex Differences in the Human Skeleton," *Evolutionary Anthropology: Issues, News, and Reviews* 29, no. 3 (2020): 108–116.

87   **"genital geometry":** Joan Roughgarden, *Evolution's Rainbow: Diversity, Gender, and Sexuality in Nature and People* (Berkeley: University of California Press, 2013), 158.

87   **"All our organs are multifunctional":** Joan Roughgarden, interview by the author, July 6, 2021.

87   **Some primatologists have gone so far:** Amy Parish, "The Evolution of the Bonobo Clitoris Through Sexual Selection," Presidential Session on the Science and Culture of the Orgasm, American Association of Anthropology, Annual Meetings, San Jose, CA, November 2006.

87   **"It does seem more logistically favorable":** Amy Parish, interview by the author, July 29, 2021.

87   **"the frontal orientation of the bonobo vulva":** Frans B. M. De Waal, "Bonobo Sex and Society," *Scientific American,* June 1, 2006.

88   **"Biology need not limit our potential":** Roughgarden, *Evolution's Rain-bow*, 180.

88   **"some scientists privately wonder":** Virginia Gewin, "A Plea for Diversity," *Nature* 422 (March 27, 2003): 368–369.

89   **"Vagina! Vagina! Vagina!":** Personal communication with documentary filmmaker Drew Denny, October 6, 2021.

Chapter 4: Protection

93   **"Some of my patients were telling me":** Ahinoam Lev-Sagie, interview by the author, April 10, 2020.

93   **"I know her very well":** "Alma," interview by the author, April 16, 2020.

94   **21 million women:** "Bacterial Vaginosis (BV) Statistics," Centers for Disease Control and Prevention, February 10, 2020, www.cdc.gov/std/bv/stats.htm/.

94   **they could revitalize the gut milieu:** Els van Nood et al., "Duodenal Infusion of Feces for Recurrent Clostridium Difficile," *The New England Journal of Medicine* 368, no. 22 (2013): 407–415.

95   **the vagina is usually dominated:** Monique Brouillette, "Decoding the Vaginal Microbiome," *Scientific American*, February 28, 2020.

95   **Nearly half a million Americans:** Fernanda C Lessa et al., "Burden of Clostridium Difficile Infection in the United States," *The New England Journal of Medicine* 372, no. 24 (2015): 2369–2370.

96   **"BV is a quality-of-life-threatening infection":** Caroline Mitchell, interview by the author, April 9, 2020.

96   **how devastating BV can be:** Jade E. Bilardi, Catriona Bradshaw, et al., "The Burden of Bacterial Vaginosis: Women's Experience of the Physical, Emotional, Sexual and Social Impact of Living with Recurrent Bacterial Vaginosis," *PloS ONE* 8, no. 9 (2013): e74378.

96   **have double the normal risk for preterm births:** Gabrielle Emanuel, "First Vaginal Bacteria Transplants in the US to Begin at Mass. General Hospital," *GBH* 89, no. 7 (2020); Herald Leitich and Herbert Kiss, "Asymptomatic Bacterial Vaginosis and Intermediate Flora as Risk Factors for Adverse Pregnancy Outcome," *Best Practice & Research: Clinical Obstetrics & Gynaecology* 21, no. 3 (2007): 375–390.

96   **higher risk of contracting HIV:** Craig R. Cohen et al., "Bacterial Vaginosis Associated with Increased Risk of Female-to-Male HIV-1 Transmission: A Prospective Cohort Analysis Among African Couples," *PloS Medicine* 9, no. 6 (2012): e1001251, 1–9; Julius Atashili et al., "Bacterial Vaginosis and HIV

Acquisition: A Meta-Analysis of Published Studies," *AIDS* (London) 22, no. 12 (2008): 1493–1501.

96 **"It's very personal":** Victoria Field, interview by the author, May 18, 2021.

97 **her microbiome had been transformed:** Ahinoam Lev-Sagie et al., "Vaginal Microbiome Transplantation in Women with Intractable Bacterial Vaginosis," *Nature Medicine* 25, no.10 (2019): 1500–1504.

97 **long-term female sex partners:** Erica L. Plummer et al., "Sexual Practices Have a Significant Impact on the Vaginal Microbiota of Women Who Have Sex with Women," *Scientific Reports* 9, no.1 (December 2019): 19749.

98 **He pointed to it as the prime culprit:** Herman Gardner and Charles Dukes, "Identification of Haemophilus Vaginalis," *Journal of Bacteriology* 81, no. 2 (1961): 277–283; "Haemophilus Vaginalis Vaginitis After Twenty-Five Years," *American Journal of Obstetrics & Gynecology.* 137, no. 3 (June 1980): 385–391.

98 **"While the disease is not a serious one":** Herman Gardner and Charles Dukes, "Haemophilus Vaginalis Vaginitis," *Central Association of Obstetricians and Gynecologists* 69, no. 5 (May 1955): 962–976.

99 **American gynecology began:** Deirdre Cooper Owens, *Medical Bondage: Race, Gender, and the Origins of American Gynecology* (Athens, GA: University of Georgia Press, 2017).

99 **in men, BV-associated communities can establish:** Marcela Zozaya et al., "Bacterial Communities in Penile Skin, Male Urethra, and Vaginas of Heterosexual Couples with and Without Bacterial Vaginosis," *Microbiome* 4 (2016): 16.

100 **"The engine of slavery rests":** Deirdre Cooper Owens, interview by the author, June 11, 2021.

100 **including false stereotypes:** Kelly M. Hoffman et al., "Racial Bias in Pain Assessment and Treatment Recommendations, and False Beliefs About Biological Differences Between Blacks and Whites" *Proceedings of the National Academy of Sciences of the United States of America* 113, no.16 (2016): 4296–4301.

100 **He claimed his patients willingly agreed:** L. L. Wall, "The Medical Ethics of Dr. J. Marion Sims: A Fresh Look at the Historical Record," *Journal of Medical Ethics,* 32, no. 6 (June 2006): 346–350.

101 **"Introducing the bent handle":** J. Marion Sims, *The Story of My Life*, ed. H. Marion-Sims (New York: D. Appleton and Company, 1884), 234.

101 **He was memorialized:** Camila Domonoske, " 'Father of Gynecology,' Who Experimented on Slaves, No Longer on Pedestal in NYC," *The Two-Way*, NPR, April 17, 2018.

104   **39 trillion or so of our closest friends:** Ed Yong, *I Contain Multitudes* (New York: HarperCollins, 2016), 10.

104   **one liquid milliliter of vaginal secretion:** William Herbert et al., *Obstetrics and Gynecology* (United States: Wolters Kluwer Health, 2013), 260.

104   **three hundred or so different species:** David A. Relman, "Learning about who we are," *Nature* 486 (2012), 194–195. DOI: 10.1038/486194a; email communication with Jacques Ravel by the author, September 30, 2021.

104   **"the body is not a thing,"** Simone de Beauvoir, *The Second Sex*, ed. Constance Borde (United Kingdom: Vintage, 2011), 46.

105   **a similar but unique microbiome in the cervix:** Chen Chen et al., "The Microbiota Continuum Along the Female Reproductive Tract and Its Relation to Uterine-Related Diseases," *Nature Communications* 8 (2017): 875.

106   **"the king of lactobacillus":** Jacques Ravel, interview by the author, February 7, 2020.

108   **Lydia E. Pinkham's Sanative Wash:** "Lydia E. Pinkham's Sanative Wash," Smithsonian National Museum of American History, www.si.edu/object/iabl-e-pinkhams-sanative-wash%3Anmah_1339291/.

108   **Lysol followed suit, marketing itself:** Andrea Tone, *Devices and Desires: A History of Contraceptives in America* (United States: Farrar, Straus and Giroux, 2002), 160–164.

108   **rarely used today:** Miranda A. Farage and Howard I. Maibach, eds., *The Vulva: Physiology and Clinical Management* (Boca Raton, FL: CRC Press, 2017), 18.

108   **nearly 1 in 5 American women still douche:** "Douching," NIH Office on Women's Health, https://www.womenshealth.gov/a-z-topics/douching/.

108   **strip the vagina of its natural protection:** Jennifer Gunter, *The Vagina Bible: The Vulva and the Vagina* (New York: Citadel Press, 2019), 104.

109   **"it's common for women":** Olga Khazan: "The Blesser's Curse," *The Atlantic*, March 2018.

110   **microbiomes generally fell:** Jacques Ravel et al., "Vaginal Microbiome of Reproductive-Age Women," *Proceedings of the National Academy of Sciences* 108, Supplement 1 (March 2011): 4680–4687.

111   **"Medicine is supposed to be unbiased":** Jessica Wells, interview by the author, April 28, 2021.

111   **a 2006 study suggested that chronic stressors:** Tonja R. Nansel, et al, "The association of psychosocial stress and bacterial vaginosis in a longitudinal cohort," *American Journal of Obstetrics and Gynecology* 194, no. 2 (2006): 381–386. DOI: 10.1016/j.ajog.2005.07.047.

112   **more than double the risk of BV:** J. F. Culhane, V. A. Rauh, and R. L. Gold-

enberg, "Stress, Bacterial Vaginosis, and the Role of Immune Processes," *Current Infectious Disease Reports* 8, no. 6 (November 2006): 459–464.

112   **"There is a budding narrative":** Jessica S. Wells et al., "The Vaginal Microbiome in U.S. Black Women: A Systematic Review," *Journal of Women's Health* (March 2020).

113   **That we could be colonized:** E. A. Miller et al., "Lactobacilli Dominance and Vaginal pH: Why Is the Human Vaginal Microbiome Unique?" *Frontiers in Microbiology* 7 (2016): 1936.

114   **"It might feel like a bit of a salve":** Willa Huston, interview by the author, March 6, 2020.

114   **a strong predictor:** Jo-Ann S. Passmore and Heather B. Jaspan, "Vaginal Microbes, Inflammation, and HIV Risk in African Women," *Lancet Infectious Diseases* 18, no. 5 (2018): 483–484.

115   **"We're just trying to think creatively":** JoAnn Passmore, interview by the author, February 23, 2021.

Chapter 5: Creation

Most of the scenes from Miriam Menkin's life, including the moment of her discovery and details about her personal life, were reconstructed by using materials from her archives at the Harvard Countway Library's Center for the History of Medicine. These include her lab notebooks, newspaper clippings, unpublished poetry, saved photos, notes for lectures, and letters to friends, colleagues, and relatives.

117   **convenient source of human subjects:** Marsh and Bonner, *The Fertility Doctor*, 72.

118   **"a delicate and frustrating business":** Sarah Rodriguez, "Watching the Watch-Glass: Miriam Menkin and One Woman's Work in Reproductive Science, 1938–1952," *Women's Studies* 44 (2015): 451–467.

118   **"She was a scientist":** Margaret Marsh, interview by the author, October 1, 2019.

119   **"the early stages of the baby":** Transcripts from interviews with journalist Loretta McLaughlin for her 1982 book *The Pill, John Rock and the Church*, Box 2, Folder 9, Countway Harvard Library.

120   **his plan to cure the scourge of infertility:** Margaret S. Marsh and Wanda Ronner, *The Fertility Doctor: John Rock and the Reproductive Revolution* (Baltimore: Johns Hopkins University Press, 2008), 131.

120   **"What a boon for the barren women":** "Conception in a Watch Glass," *The New England Journal of Medicine* 217 (October 1937): 678.

121 **"She wasn't just somebody's gofer"**: Marsh, interview, October 1, 2019.

121 **"Dr. Rock was never there"**: McLaughlin transcripts, Box 2, Folder 8.

121 **She hated to take her $1.25 an hour**: McLaughlin transcripts, Box 2, Folder 9.

122 **"an *in vivo* specimen"**: Rachel E. Gross, "The Female Scientist Who Changed Human Fertility Forever," *BBC Future*, January 5, 2020, www.bbc.com/future/article/20200103-the-female-scientist-who-changed-human-fertility-forever.

122 **"It was just something I had to do"**: McLaughlin transcripts, Box 2, Folder 8.

122 **she would run home**: McLaughlin transcripts, Box 2, Folder 8.

122 **"A real country doctor"**: McLaughlin transcripts, Box 2, Folder 10.

123 **a degree in histology and comparative anatomy**: Rodriguez, "Watching the Watch-Glass."

123 **"I really was nobody"**: McLaughlin transcripts, Box 2, Folder 8.

124 **fertilized rabbit eggs in vitro**: Miriam F. Menkin and John Rock, "In Vitro Fertilization and Cleavage of Human Ovarian Eggs," *American Journal of Obstetrics and Gynecology* 55, no. 3 (1948): 440–452.

125 **"where woman would be self-sufficient"**: Marsh and Ronner, *Fertility Doctor*, 141.

125 **"human children being brought into the world"**: William L. Laurence, "Life Is Generated in Scientist's Tube," *New York Times*, March 27, 1936.

125 **"Pincus did not seek publicity"**: Wanda Ronner and Margaret Marsh, *The Pursuit of Parenthood: Reproductive Technology from Test-Tube Babies to Uterus Transplants* (Baltimore: Johns Hopkins University Press, 2019), 19.

125 **In 1937, he was denied tenure**: Marsh and Ronner, *Fertility Doctor*, 141.

126 **the political tide had changed**: David A. Valone, "The Changing Moral Landscape of Human Reproduction: Two Moments in the History of In Vitro Fertilization," *The Mount Sinai Journal of Medicine* 65, no. 3 (May 1998).

126 **"scientific affront to womanhood"**: "On Nature's Heels," *Time*, August 14, 1944.

127 **"I was so exhausted and drowsy"**: "World's First *In Vitro* Fertilized Egg," *Inside AHC*. Monthly bulletin published for employees of the Affiliated Hospitals Center, February 1979, page 3.

127 **"nearly dropped dead"**: McLaughlin transcripts, Box 2, Folder 8.

128 **"the first miscarriage in a test tube"**: "Biologist Miriam Menkin Recalls Pioneer Efforts," *Morning Call* (Allentown, PA), July 30, 1978.

128 **"our pride and joy"**: McLaughlin transcripts, Box 2, Folder 8.

128 **reported their findings in a brief initial report:** John Rock and Miriam F. Menkin, "In Vitro Fertilization and Cleavage of Human Ovarian Eggs," *Science* 100, no. 2588 (1944): 105–107. DOI: 10.1126/science.100.2588.105.

128 **"Whether we shall be able to fertilize human egg cells":** Robert S. Bird, "A Human Ovum Is Fertilized in Test Tube for the First Time," Menkin Papers, undated Boston newspaper clipping August 1944.

128 **"Dr. Rock's experiment is the first ray of hope":** "Momentous Conception: A New Human Life Formed Under the Microscope," *Science Illustrated*, September 1944, 49.

129 **her husband lost his job:** McLaughlin transcripts, Box 2, Folder 8.

129 **"That was really very sad":** McLaughlin transcripts, Box 2, Folder 98.

129 **"rape in vitro":** Marsh and Ronner, *Fertility Doctor*, 109.

129 **"My life's ambition was to have a chance to repeat it":** McLaughlin transcripts, Box 2, Folder 8.

129 **"the instrument of the Devil":** Menkin papers, Box 5, Folder 6.

130 **"They are the only cells":** Scott Pitnick, interview by the author, November 17, 2019.

130 **"Sperm is a drop of brain":** Edward Dolnick, *The Seeds of Life: From Aristotle to da Vinci, from Sharks' Teeth to Frogs' Pants, the Long and Strange Quest to Discover Where Babies Come From* (New York: Basic Books, 2017), 45.

130 **"a dormant bride awaiting":** Gerald Schatten and Heide Schatten, "The Energetic Egg," *Medical World News* 23 (January 23, 1984): 51.

130 **"imagery keeps alive some of the hoariest":** Emily Martin, "The Egg and the Sperm," *Signs* 16, no. 3 (Spring 1991): 485–501.

130 **"sinfully defiling myself . . . addle-pated":** Dolnick, *Seeds of Life*, 114–121.

130 **he had produced his sample by "conjugal coitus":** Dolnick, *Seeds of Life*, 114.

131 **up to 30 millimeters:** Kenneth Saladin, *Human Anatomy*, 5th ed. (New York: McGraw-Hill Education, 2017), 1093.

132 **"the goal of achieving fertilization *in vitro* in a mammal was stymied":** Scott Pitnick, Mariana F. Wolfner, and Steve Dorus, "Post-Ejaculatory Modifications to Sperm (PEMS)," *Biological reviews of the Cambridge Philosophical Society* 95, no. 2 (April 2020): 365–392.

132 **"I don't think we understand ":** Kurt Barnhart, interview by the author, October 23, 2019.

133 **they flail far too weakly:** Martin, "The Egg."

133 **"If I had to wager":** Pitnick, interview by the author, 2019.

135   **"When I observed the ovary"**: Dolnick, *Seeds of Life,* 256.

136   **his microscope and a smear of sea-urchin jizz**: Dolnick, *Seeds of Life,* 262.

136   **"arises to completion"**: Dolnick, *Seeds of Life,* 262.

137   **"with a single brilliant stroke, illuminated the field"** Susan G. Ernst, "A Century of Sea Urchin Development." *American Zoologist* 37, no. 3 (1997): 250–259, www.jstor.org/stable/3883920.

137   **"The nuclei of the two germ-cells"**: E. B. Wilson, *An Atlas of the Fertilization and Karyokinesis of the Ovum* (New York and London: Macmillan, 1895).

137   **"When you think how small this egg is"**: Menkin papers, Box 1, Folder 53.

137   **"It took someone who was"**: Sarah Rodriguez, interview by the author, January 28, 2019.

138   **"a dirty trick"**: McLaughlin transcripts, Box 2, Folder 8.

138   **the complete version of her first brief IVF report**: Menkin and Rock, "In Vitro fertilization and Cleavage."

139   **at Rock's urging, listed as first author**: Marsh and Ronner, *Fertility Doctor,* 109.

139   **"I always felt that I should pay Dr. Rock"**: Menkin Papers, Box 2, Folder 9.

139   **Valy sat across the dinner table**: Menkin Papers, Box 2, Folder 62.

140   **"I believed it would create too much of a trauma"**: Menkin Papers, Box 2, Folder 62.

140   **"The joy of new-found FREEDOM:"** Menkin Papers, Box 2, Folder 62.

141   **hadn't had a regular salary**: McLaughlin transcripts, Box 2, Folder 8.

142   **"Mrs. Menkin, why don't you come back to me"**: McLaughlin transcripts, Box 2, Folder 8.

143   **"She was just this humble, humble person"**: David Albertini, interview by the author, May 11, 2020.

143   **8 million babies**: Georg Griesinger, "Is Progress in Clinical Reproductive Medicine Happening Fast Enough?" *Upsala Journal of Medical Sciences* 125, no. 2 (2020), 65–67.

143   **"youth (under 35)"**: Anne Taylor Fleming, "New Frontiers in Conception," *New York Times,* July 20, 1980.

143   **"profoundly revolutionary technology"**: Rene Almeling, interview by the author, May 13, 2021.

145   **"It was literally just a footnote"**: Elizabeth Carr, interview by the author, May 12, 2021.

146   **Carnegie took the lead**: Adrianne Noe, "The Human Embryo Collection"

in *Centennial History of the Carnegie Institute of Washington*, ed. Jane Maienshein (Cambridge University Press, 2004), 35.

146 **"the 3rd stage cellular reaction by division":** Valone, "Changing Moral Landscape," 170.

147 **named after the doctor who procured them:** Noe, "Human Embryo Collection," 36.

147 **"This is pretty much as early":** Elizabeth Lockett, interview by the author, October 9, 2019.

Chapter 6: Power

149 **"Why do we age?":** Jon Tilly, interview by the author, January 22, 2021.

150 **Eggs don't wither away at random:** Jon Tilly, "The Genes of Cell Death and Cellular Susceptibility to Apoptosis in the Ovary: A Hypothesis," *Cell Death & Differentiation* 4 (1997): 180–187.

150 **Their death is genetically pre-programmed:** Jon Tilly, "Commuting the Death Sentence: How Oocytes Strive to Survive," *Nature Reviews Molecular Cell Biology* 2 (2001): 838–848.

150 **Starting before birth:** Jerome F. Strauss, Robert L. Barbieri, and Samuel S. C. Yen, *Yen & Jaffe's Reproductive Endocrinology: Physiology, Pathophysiology, and Clinical Management,* 8th ed, (Philadelphia: Elsevier, 2019), 172.

151 **rate of up to one-third of the total oocyte pool:** Michael Lemonick, "Of Mice and Menopause," *Time,* March 22, 2004.

151 **men produce more than a thousand sperm:** Christine Dell'Amore, "How a Man Produces 1,500 Sperm a Second," *National Geographic*, March 19, 2010.

152 **increasing the chances of genetic mutations:** Arslan A. Zaidi et al, "Bottleneck and selection in the germline and maternal age influence transmission of mitochondrial DNA in human pedigrees," *Proceedings of the National Academy of Sciences*, December 2019, 116 (50) 25172–25178. DOI: 10.1073/pnas.1906331116.

152 **Maybe women weren't simply running out:** Evelyn Telfer and Jonathan Tilly, "Purification of Germline Stem Cells from Adult Mammalian Ovaries: A Step Closer Towards Control of the Female Biological Clock?" *Molecular Human Reproduction* 15, no. 7 (July 2009): 393–398.

155 **women might have new options:** Alvin Powell, "Examining Cell Death, Researchers Explode Belief About Life," *The Harvard Gazette,* March 25, 2004.

155 **the paper came out in the journal *Nature*:** Joshua Johnson et al., "Germ-

line Stem Cells and Follicular Renewal in the Postnatal Mammalian Ovary," *Nature* 428, no. 6979 (March 2004): 145–150.

156 **"women could essentially grow back":** Emma Croager, "Egg-Citing Fertility Finding," *Nature Reviews Molecular Cell Biology* 5, no. 256 (2004), www.nature.com/articles/nrm1376.

156 **"could be the most significant advance":** Roger Highfield, "Scientists Find a Way to Beat the Menopause," *The Telegraph,* March 11, 2004.

156 **"The ability to make more eggs":** Natalie Angier, "Scientists Find Indications That Ovaries May Be Replenished," *New York Times*, March 10, 2004.

156 **"This is the way of science":** Evelyn Telfer, "Germline stem Cells in the Postnatal Mammalian Ovary: A Phenomenon of Prosimian Primates and Mice?" *Reproductive Biology and Endocrinology* 2, no. 24 (2004), www.ncbi.nlm.nih.gov/pmc/articles/PMC434530/.

156 **"Phones were ringing off the hook":** Kendall Powell, "Going Against the Grain," *PloS Biology* 5, no. 12 (2007): e338. DOI: 10.1371/journal.pbio.0050338.

157 **he was "utterly blasted":** Joshua Johnson, interview by the author, May 11, 2020.

159 **the Study of Women's Health Across the Nation:** Nanette Santoro, "The SWAN Song: Study of Women's Health Across the Nation's Recurring Themes," *Obstetrics and Gynecology Clinics of North America* 38, no. 3 (September 2011): 417–423.

159 **"It's an awful lot of work":** Dori Woods, interview with the author, May 12, 2020.

160 **testosterone plays a key role:** Rebecca Jordan-Young and Katrina Karkazis, *Testosterone: An Unauthorized Biography* (Cambridge, MA: Harvard University Press, 2019): 38–39.

161 **"they are essential for the oocyte's ability":** John Eppig, interview by the author, May 8, 2020.

161 **a crackling network of communication:** Stephen S. Hall, "The Good Egg," *Discover*, May 28, 2004.

163 **the ultimate goal of building an "artificial ovary":** T. Akahori, D. C. Woods, and J. L. Tilly, "Female Fertility Preservation through Stem Cell-based Ovarian Tissue Reconstitution In Vitro and Ovarian Regeneration In Vivo," *Clinical Medicine Insights: Reproductive Health* 13 (May 2019): 1–10.

164 **Chinese team reported:** Q. Wu et al., "CARM1 Is Required in Embryonic Stem Cells to Maintain Pluripotency and Resist Differentiation," *Stem Cells* 27, no. 11 (November 2009): 2637–2645.

165 **"Nature had her mind on the job"**: Dolnick, *Seeds of Life*, 99.

166 **Twenty-three-year-old Julia Omberg**: Thomas Schlich, "Cutting the Body to Cure the Mind," *The Lancet Psychiatry* 2, no. 5, (May 2015): 390–392.

166 **"If I could but divest her"**: Battey, Robert. "Normal Ovariotomy," *Atlanta Medical and Surgical Journal* 10, no. 6 (September 1872), collections.nlm.nih.gov /ext/dw/66970270R/PDF/66970270R.pdf/.

166 **she never had her period again**: Lawrence D. Longo, "The Rise and Fall of Battey's Operation: A Fashion in Surgery," *Bulletin of the History of Medicine* 53, no. 2 (Summer 1979): 244–267.

166 **scientists had not yet linked them definitively**: Longo, "Battey's Operation," 265–266.

167 **the mortality rate was close to 1 in 3**: Longo, "Battey's Operation," 252.

167 **named the procedure after Battey**: Sally Frampton, *Belly-Rippers: Surgical Innovation and the Ovariotomy Controversy* (Palgrave Macmillan, 2018), 119, library.oapen.org/bitstream/20.500.12657/22950/1/1007211.pdf

167 **one doctor estimating that 150,000 women**: Longo, "Battey's Operation," 253.

167 **paving the way for a renaissance**: Sir T. Spencer Wells. "Modern Abdominal Surgery, with an Appendix on the Castration of Women," The Bradshaw Lecture, delivered December 18, 1890, London: J. & A. Churchill (1891).

167 **"It was like the discovery"**: Sir T. Spencer Wells, "Castration in Mental and Nervous Diseases," *The American Journal of the Medical Sciences* (J. B. Lippincott, Company, 1886), 456.

167 **"I decide, in such cases"**: Longo, "Battey's Operation," 250.

167 **"the surgeon can console himself"**: Longo, "Battey's Operation," 259.

167 **his overuse of the operation**: Longo, "Battey's Operation," 261.

168 **doctors soon raised concerns**: Longo, "Battey's Operation," 263.

168 **"unwomanly pilosity on the chin"**: Eugen Steinach and Josef Loebel, *Sex and Life* (New York: The Viking Press, 1940), 49.

168 **"spaying," "unsexing," or "castration of women"**: Longo, "Battey's Operation," 244.

168 **the ovaries' role in reproductive physiology**: Longo, "Battey's Operation," 265–266.

169 **physiologist Ernest Starling named them hormones**: Randi Epstein, *Aroused: The History of Hormones and How They Control Just About Everything*, (New York W. W. Norton & Company, 2018), 28–29.

169 **"a war-ridden world"**: Buren Thorne, "The Craze for Rejuvenation," *New York Times*, June 4, 1922, 54.

169   **the wild bullock into the patient ox:** Steinach and Loebel, *Sex and Life*, 3.

169   **"the very citadels of sex":** Steinach and Loebel, *Sex and Life*, 61.

171   **notorious gland peddlers:** John R. Herman, "Rejuvenation: Brown-Séquard to Brinkley: Monkey Glands to Goat Glands," *New York State Journal of Medicine* 82, no. 2 (1982): 1731–1739.

172   **cause sperm-producing cells to shrivel up:** "Doctor Undergoes Steinach Operation: Dr. David T. Marshall Submits to Knife to Relieve High Blood Pressure," *New York Times*, 1923.

172   **"Steinaching" rocketed in popularity:** Chandak Sengoopta, " 'Dr. Steinach Coming to Make Old Young!': Sex Glands, Vasectomy and the Quest for Rejuvenation in the Roaring Twenties," *Endeavor* 27, no. 3 (September 2003).

173   **"convert grandmothers into debutantes":** "Doctor Offers Cure for Age," *Toledo Blade*, August 30, 1923.

173   **she recommended that Germany "Steinach" its population:** Associated Press, "Mrs. Atherton Causes Amusement in Berlin—Newspapers Ridicule Her Suggestion for Rejuvenation of All Germany's 'Supermen,'" *New York Times*, April 6, 1924.

174   **they named it for one conspicuous effect:** Edgar Allen and Edward A. Doisy, "An Ovarian Hormone," *The Journal of the American Medical Association* 81 (1923): 819–821.

174   **One of its first uses:** Richard J. Santen and Evan Simpson, "History of Estrogen: Its Purification, Structure, Synthesis, Biologic Actions, and Clinical Implications," *Endocrinology* 160, no. 3 (March 2019): 605–625.

174   **a method recently developed:** Arthur T. Hertig, "Allen and Doisy's 'An Ovarian Hormone,'" *The Journal of the American Medical Association* 250, no. 19 (November 18, 1983): 2684–2688.

175   **a convenient way of to market:** Nelly Oudshoorn, *Beyond the Natural Body: An Archeology of Sex Hormones* (United Kingdom: Taylor & Francis, 2003), 95.

175   **pharmaceutical companies first had to sell menopause:** Frances B. McCrea, "The Politics of Menopause: The 'Discovery' of a Deficiency Disease," *Social Problems* 31, no. 1 (October 1983): 111–123.

175   **"you have a new assurance":** "How Women Over 35 Can Look Younger," *LIFE*, January 23, 1950, 40.

176   **"The unpalatable truth must be faced":** Robert A. Wilson and Thelma A. Wilson, "The Fate of the Nontreated Postmenopausal Woman: A Plea for the Maintenance of Adequate Oestrogen From Puberty to the Grave," *Journal of the American Geriatric Society*, no. 11 (1963), 347–62.

176 **Wilson was paid by the companies:** Morton Mintz, *The Pill: An Alarming Report* (Boston: Beacon Press, 1970), 30.

176 **"acceptable feminine appearance":** T. S. Cairns and W. De Villiers, "Vaginoplasty," *South African Medical Journal* 57, no. 2 (1980): 52.

177 **promote brain development:** Matthew C. S. Denley et al., "Estradiol and the Development of the Cerebral Cortex: An Unexpected Role?" *Frontiers in Neuroscience* 12 (2018): 245. DOI: 10.3389/fnins.2018.00245.

177 **"sex hormone" be changed to "growth hormone":** Anne Fausto-Sterling, *Sexing the Body: Gender Politics and the Construction of Sexuality.* (New York: Basic Books, 2008), 193.

177 **estrogen produced by the testes:** Michael Schulster, Aaron M. Bernie, Ranjith Ramasamy, "The Role of Estradiol in Male Reproductive Function," *Asian Journal of Andrology* 18, no. 3 (2016): 435–440. DOI: 10.4103/1008-682X .173932.

177 **brain development:** Jordan-Young and Karkazis, *Testosterone,* 38–39. For a full discussion of the wide-ranging, messy effects of estrogen and testosterone—and testosterone's role in ovulation and ovarian health, see Young and Karkazis, *Testosterone.* For an examination of how estrogen and testosterone became known as sex hormones, see *Beyond the Natural Body: An Archeology of Sex Hormones* by Nelly Oudshoorn.

178 **estrogen sales in the United States quadrupled:** McCrea, "Politics of Menopause," 114.

178 **hormone therapy did not necessarily protect women:** J. E. Manson et al. "Menopausal Hormone Therapy and Health Outcomes During the Intervention and Extended Poststopping Phases of the Women's Health Initiative Randomized Trials," *The Journal of the American Medical Association* 310, no. 13 (2013): 1353–1368.

179 **"A fifty-four-year-old ovary should not":** Jen Gunter, interview by the author, July 9, 2021.

179 **they pump out trace amounts of estradiol:** Strauss, Barbieri, and Yen, *Reproductive Endocrinology,* 204; Jen Gunter, *The Menopause Manifesto* (New York: Citadel Press, 2021), 31.

180 **He's since moved on:** Clare Wilson, "Ovary Freezing Offers a Drug-Free Way to Tame Menopause," *New Scientist,* December 30, 2015.

180 **"We can beat the biologic clock":** Sherman Silber, interview by the author, May 15, 2020.

180 **at least eleven women had paid:** Charlotte Hayward, "Concerns over New

'Menopause Delay' Procedure," *BBC News*, January 28, 2020, www.bbc.com/news/health-51269237.

182 **"These are the cells that don't exist":** Dori Woods, interview by the author, January 22, 2021.

183 **she went on to coauthor several papers with Tilly:** Jonathan Tilly and Evelyn Telfer, "Purification of Germline Stem Cells from Adult Mammalian Ovaries: A Step Closer Towards Control of the Female Biological Clock?" *Molecular Human Reproduction* 15, no. 7 (2009): 393–398. DOI: 10.1093/molehr/gap036.

Chapter 7: Regeneration

Portions of this chapter were previously published in the *New York Times* on April 27, 2021, under the title, "They Call It a 'Women's Disease.' She Wants to Redefine It."

185 **Could it be a cyst?:** Most of the scene about Linda's cancer experience is reconstructed through multiple interviews with Linda Griffith, her husband, Doug Lauffenburger, and her friends and colleagues who saw her through this year.

185 **diagnosed with triple-negative breast cancer:** Gina Kolata, "Cancer Fight: Unclear Tests for New Drug," *New York Times,* April 19, 2010.

187 **"We transformed our lab meetings":** Nicole Doyle, interview by the author, September 18, 2020.

187 **analyzing peritoneal fluid:** Amanda Schaffer, "The Practical Activist," *MIT Technology Review,* August 19, 2014.

187 **the first study to propose a way to categorize endo patients:** M. T. Beste et al., "Molecular Network Analysis of Endometriosis Reveals a Role for C-Jun-Regulated Macrophage Activation," *Science Translational Medicine* 5, no. 6 (February 2014): 222ra16.

187 **"I viewed it as a terrible thing":** Doug Lauffenburger, interview by the author, August 24, 2020.

188 **"I was working on all the things":** Video recording of NIH Meeting on Menstruation: Science and Society, September 20, 2018, videocast.nih.gov/watch=28461/.

188 **"there was nothing we couldn't do":** Susan Berthelot, interview with the author, September 1, 2020.

189 **"rejecting her femininity":** Linda Griffith, interview with the author, July 20, 2020.

190 **an iconic creature:** C. Vacanti et al., "Tissue Engineered Growth of New Cartilage in the Shape of a Human Ear Using Synthetic Polymers Seeded with Chondrocytes," *MRS Proceedings* 25 (1991): 367.

190 **a human ear-shaped scaffold:** W. S. Kim et al., "Cartilage Engineered in Predetermined Shapes Employing Cell Transplantation on Synthetic Biodegradable Polymers," *Plastic and Reconstructive Surgery*. 94, no. 2 (August 1994): 233–237.

190 **"the career woman's disease":** Carolyn Carpan, "Representations of Endometriosis in the Popular Press: 'The Career Woman's Disease,'" *Atlantis* 27, no. 2 Health Panic and Women's Health (2003).

190 **"underweight, overanxious, intelligent":** S. L. Darrow et al., "Sexual Activity, Contraception, and Reproductive Factors in Predicting Endometriosis," *American Journal of Epidemiology* 140 (1994): 500–509.

190 **They commonly prescribed marriage and pregnancy:** J. W. McArthur, "The effect of Pregnancy Upon Endometriosis," *Obstetrical & Gynecological Survey* 20, no. 5 (October 1965): 709–733.

191 **has been roundly refuted:** Brigitte Leeners et al., "The Effect of Pregnancy on Endometriosis—Facts or Fiction?" *Human Reproduction Update* 24, no. 3 (May–June 2018): 290–299.

191 **doctors still recommend pregnancy:** Kate Young, "Infertility Is an Issue for Some Women with Endometriosis, But It's Not the Whole Story," *The Guardian*, December 2, 2018.

191 **woman was wetter and spongier:** Helen King, *Hippocrates' Woman: Reading the Female Body in Ancient Greece* (London: Routledge, 1998), 28.

192 **it was a medical diagnosis:** Helen King, "Once Upon a Text," in *Hysteria Beyond Freud*, eds. Sander L. Gilman et al. (Berkeley: University of California Press, 1993), 14.

193 **"It's a very useful way":** Helen King, interview by the author, October 29, 2020.

193 **A Frenchwoman leans backward:** André Brouillet, *A Clinical Lesson at the Salpêtrière*, painting, 1887.

194 **one that lived in the brain:** Gilman et al., *Hysteria Beyond Freud* (Berkeley: University of California Press, 1993), 13.

194 **he would demonstrate a hysterical attack:** Barry Stephenson, "Charcot's Theatre of Hysteria," *Journal of Ritual Studies* 15, no. 1 (2001): 27–37.

194 **hysterical episodes could be triggered:** Gilman et al., *Hysteria Beyond Freud*, 307.

195 **a medical student named Sigmund Freud:** Elaine Showalter, "Hysteria, Feminism, and Gender," in Gilman et al., *Hysteria Beyond Freud*, 314.

195 **working in the neurology lab:** L. C. Triarhou, "Exploring the Mind with a Microscope: Freud's Beginnings in Neurobiology," *Hellenic Journal of Psychology* 6, no. (2009): 1–13.

195 **"But, my dear sir, how can you talk":** Sigmund Freud, "An Autobiographical Study," *The Standard Edition of the Complete Psychological Works of Sigmund Freud* (London: Hogarth Press, 1959), 15. (Translation from German by James Strachey.)

195 **"hysteria behaves as though anatomy did not exist":** Sigmund Freud, "Some Points for a Comparative Study of Organic and Hysterical Motor Paralyses," (1893) Standard Edition 1: 157–172, 1966.

197 **"Hysteria dressed up in modern garb":** Maya Dusenbery, *Doing Harm: The Truth About How Bad Medicine and Lazy Science Leave Women Dismissed, Misdiagnosed, and Sick,* First edition (New York: HarperOne, 2018), 78.

198 **"You were probably molested as a child":** Abby Norman, *Ask Me About My Uterus: A Quest to Make Doctors Believe in Women's Pain* (New York: Nation Books, 2018), 128.

198 **"the new hysteria":** Kate Young, Maggie Kirkman, and Jane Fisher, "Is Endometriosis the New Hysteria? Modern Day Implications for Medicine's Historical Construction of Women and Their Bodies," Paper presented at the Australian Society for Psychosocial Obstetrics & Gynaecology 41st Annual Scientific Meeting in Melbourne, Australia, August 2015; Kate Young, Jane Fisher, and Maggie Kirkman, "Do Mad People Get Endo or Does Endo Make You Mad?: Clinicians' Discursive Constructions of Medicine and Women with Endometriosis," *Feminism & Psychology* 29, no. 3 (August 2019): 337–356.

199 **"Bountiful amounts of energy":** Pardis Sabetti, interview by the author, August 26, 2020.

199 **"Like a thunderstorm":** Steven Tannenbaum, interview by the author, September 13, 2020.

200 **"The hellmouth will open at the dinner":** Linda Griffith, interview by the author, August 25, 2020.

200 **More than half of women who undergo hysterectomies:** B. Rizk et al., "Recurrence of Endometriosis After Hysterectomy," *Facts, Views, and Vision in Obstetrics and Gynaecology* 6, no. 4 (2014): 219–227.

201 **"I have a chronic disease called endometriosis":** Recording of Women in Science and Engineering Luncheon, 2007, vimeo.com/449799275/3c42a9ee94/.

201 **"Thank God for Dr. Linda Griffith":** Recording of Padma Lakshmi speech at launch of CGR, www.youtube.com/watch?v=WKnDjTUnKEA/.

203   **"How the body can coordinate that":** Hilary Critchley, interview by the author, August 25, 2020.

203   **84 species—1.6 percent of all placental mammals:** H.O.D. Critchley et al., "Menstruation: Science and Society," *American Journal of Obstetrics and Gynecology* 223, no. 5 (November 2020): 624–664.

203   **The Cairo spiny mouse:** N. Bellofiore et al., "A Missing Piece: The Spiny Mouse and the Puzzle of Menstruating Species," *Journal of Molecular Endocrinology* 61, no. 1 (July 2018): R25–R41.

204   **menstruating women sweated toxins:** Shreya Dasgupta, "Why Do Women Have Periods When Most Animals Don't?" *BBC Earth*, April 20, 2015, www .bbc.com/earth/story/20150420-why-do-women-have-periods.

204   **"defend against pathogens":** Margie Profet, "Menstruation as a Defense Against Pathogens Transported by Sperm," *The Quarterly Review of Biology* 68, n. 3 (September 1993): 335–386.

204   **motherhood isn't all warm and fuzzy:** Deena Emera et al., "The Evolution of Menstruation: A New Model for Genetic Assimilation: Explaining Molecular Origins of Maternal Responses to Fetal Invasiveness," *BioEssays* 34, no. 1 (2012): 26–35.

204   **"a little bottlebrush that has eyes and a tail":** Nadia Bellofiore, interview by the author, December 28, 2020.

205   **the differentiation of the uterine lining:** Critchley et al., "Menstruation: Science and Society," 624–664.

205   **"sort of like a standing army":** Günter Wagner, interview by the author, December 9, 2020.

206   **they're Type-A planners:** Laura Catalini and Jens Fedder, "Characteristics of the Endometrium in Menstruating Species: Lessons Learned from the Animal Kingdom," *Biology of Reproduction* 102, no. 6 (June 2020): 1160–1169.

206   **whether a prospective embryo should live or die:** Nick S. Macklon and Jan J. Brosens, "The Human Endometrium as a Sensor of Embryo Quality," *Biology of Reproduction* 91, no. 4 (October 2014), 1–8.

207   **"I'm increasingly thinking":** Kate Clancy, interview by the author, September 14, 2020.

207   **"It is likely that the mechanisms deployed":** Critchley et al., "Menstruation: Science and Society," 624–664.

207   **"chocolate cysts" attached to their ovaries:** John A. Sampson, "The Development of the Implantation Theory for the Origin of Peritoneal Endometriosis," *American Journal of Obstetrics and Gynecology* 40, no. 4 (October 1940): 549–557.

208 **"affect women in a most valuable period of their lives"**: John A. Sampson, "Perforating Hemorrhagic (Chocolate) Cysts of the Ovary," *Archives of Surgery* 3 (September 1921): 245–323.

208 **"It remains a riddle"**: Adi E. Dastur and P. D. Tank, "John A Sampson and the Origins of Endometriosis," *Journal of Obstetrics and Gynaecology of India* 60, no. 4 (2010): 299–300.

208 **"Nature (since the beginning of time)"**: Clayton T. Beecham, "Surgical Treatment of Endometriosis with Special Reference to Conservative Surgery in Young Women," *The Journal of the American Medical Association* 139, no. 15 (1949): 971.

209 **"It's a variation on a theme"**: Linda Giudice, interview by the author, August 19, 2020.

209 **"It is a systemic disease of inflammation"**: Hugh S. Taylor, Alexander M. Kotlyar, and Valerie A. Flores, "Endometriosis Is a Chronic Systemic Disease: Clinical Challenges and Novel Innovations," *Lancet* 27, no. 397 (February 2021): 839–852.

209 **"We need to address this"**: Elise Courtois, interview by the author, March 25, 2021.

210 **a simple diagnostic test for endometriosis**: A. Nayyar et al., "Menstrual Effluent Provides a Novel Diagnostic Window on the Pathogenesis of Endometriosis," *Frontiers in Reproductive Health* 2, no. 3 (2020).

210 **"I don't think that's the whole story"**: Peter Gregersen, interview by the author, August 12, 2020.

211 **"It's not at all just a woman's problem"**: Kevin Osteen, interview by the author, October 8, 2020.

212 **Maisha Johnson was in bed**: Maisha Johnson, "I'm Black. I Have Endometriosis—and Here's Why My Race Matters," *Healthline*, July 10, 2019, www.healthline.com/health/endometriosis/endo-race-matters

212 **"If I'm seen as a woman who's prone to hysteria"**: Maisha Johnson, interview by the author, May 20, 2021.

213 **hours in the waiting room**: Jaipreet Virdi, "Getting the Measure of Pain," Wellcome Collection, August 15, 2019, wellcomecollection.org/articles/XTg6QxAAACIAP5f7/.

213 **"they were looking at me like I was a drug seeker"**: Jaipreet Virdi, interview by the author, May 11, 2021.

213 **doctors often lack knowledge**: J. Obedin-Maliver et al., "Lesbian, Gay, Bisexual, and Transgender–Related Content in Undergraduate Medical

Education," *The Journal of the American Medical Association* 306, no. 9 (2011): 971–977; Pauline W. Chen, "Medical Schools Neglect Gay and Gender Issues," *New York Times* Well Blog, November 10, 2016, well.blogs.nytimes.com/2011 /11/10/medical-schools-teach-little-about-gay-health-issues/.

213   **"Many clinicians are unable to disentangle":** Emily Lipstein, "Treating Endometriosis as a Women's Disease Hurts Patients of All Genders," *Vice*, November 11, 2020.

214   **he felt sure that he was a boy:** "Living with Endometriosis as a Transgender Patient," Video interview with *NowThis News*, October 18, 2018, www .youtube.com/watch?v=eczxr0bYAxw

214   **had his ovaries and uterus removed:** Alexandra Stovicek, "He Is 1 in 10: A Trans Man Shares What Life Is Like with Endometriosis," EndoFound, June 4, 2018 www.endofound.org/he-is-1-in-10-a-trans-man-shares-what -life-is-like-with-endometriosis.

214   **"As a girl, they thought":** Cori Smith, interview by the author, May 16, 2021.

Chapter 8: Beauty

This chapter drew heavily on the histories provided by three exemplary books: *How Sex Changed,* by historian Joanne Meyerowitz; *Bodies in Doubt,* by historian Elizabeth Reis; and *Transgender History,* by historian Susan Stryker.

217   **"You don't want a vagina like this":** Marci Bowers, in P. J. Raval and Jay Hodges, *Trinidad: Transgender Frontier,* documentary film, New Day Films, 2007.

218   **"It really is a continuum":** Marci Bowers, interview with the author, August 23, 2019.

219   **"the Georgia O'Keeffe of genitalia":** Raval and Hodges, *Trinidad,* 2007.

219   **"I was never a boy's boy":** Raval and Hodges, *Trinidad,* 2007.

219   **would satisfy the feelings she had:** "Gender Identity," *Oprah Winfrey Show,* September 28, 2007, www.oprah.com/oprahshow/gender-identity/9.

220   **she could hardly look in the mirror:** "Gender Identity," *Oprah.*

220   **"What patients really experience":** Lisa Capretto, "Inside the Practice of a Doctor Who Has Performed 1,500 Gender Reassignment Surgeries," *Huffington Post,* April 1, 2015, www.huffpost.com/entry/marci-bowers-gender -reassignment-transgender_n_6980782/.

221   **"and that is hope":** Marci Bowers, "Converging Identities in a Chang-

ing World," TEDxPaloAlto, May 4, 2017, www.youtube.com/watch?v=fdN
M2rFfVFY.

221 **"This is my one shot":** Roxanne Euber, interview by the author, June 2, 2019.

226 **set the standard for what a vagina should look and feel like:** Eric Plemons, "Anatomical Authorities: On the Epistemological Exclusion of Trans-
Surgical Patients," *Medical Anthropology* 34, no. 5 (2015): 425–441.

227 **"We want to keep the Montreal edge":** Pierre Brassard, interview with the
author, September 13, 2019.

227 **not only aesthetically pleasing but fully functional:** Plemons, Eric. "It Is
as It Does: Genital Form and Function in Sex Reassignment Surgery," *Journal
of Medical Humanities* 35, no. 1 (2014): 37–55.

228 **they can use a loop of peritoneum:** Megan Molteni, "A Patient Gets the New
Transgender Surgery She Helped Invent," *Wired*, September 11, 2017, www.wired.
com/story/a-patient-gets-the-new-transgender-surgery-she-helped-invent/.

228 **studies suggest that this microbial mixture:** K. D. Birse et al., "The
Neovaginal Microbiome of Transgender Women Post-Gender Reassignment
Surgery," *Microbiome* 8, no. 61 (2020).

229 **"EX-GI BECOMES BLONDE BEAUTY":** Joanne Meyerowitz, *How Sex Changed:
A History of Transsexuality in the United States* (Cambridge: Harvard University
Press, 2002), 62.

231 **"the patient had one final ardent wish":** Meyerowitz, *How Sex Changed*, 61.

233 **spreading his glandular gospel:** "Doctor Undergoes Steinach Operation:
Dr. David T. Marshall Submits to Knife to Relieve High Blood Pressure," *New
York Times*, 1923.

233 **the father of transgender medicine:** C. Wolf-Gould, "History of Transgender Medicine in the United States," in *The SAGE Encyclopedia of LGBTQ
Studies*, ed A. Goldberg (SAGE Publications, Inc., 2016), 508–512.

233 **"To me it is just a matter of relieving":** Tom Buckley, "The Transsexual
Operation," *Esquire* (April 1967), 111–116.

234 **sex was like a pointillism painting:** Harry Benjamin, *The Transsexual Phenomenon* (New York: Ace Pub. Co, 1966), 4.

236 **Denmark turned down more than four hundred appeals:** Meyerowitz,
How Sex Changed, 132.

236 **an obscure group of laws:** Susan Stryker, *Transgender History* (Berkeley, CA:
Seal Press, 2008), 44-45.

237 **"I was just a glob of aching flesh":** Meyerowitz, *How Sex Changed*, 147.

238 **"how well they lend themselves":** Elizabeth Reis, *Bodies in Doubt: An*

American History of Intersex, Second Edition (Baltimore: Johns Hopkins University Press, 2021), 114.

238 **"You can dig a hole, but you can't build a pole":** Melissa Hendricks, "Is It a Boy or a Girl?," *John Hopkins Magazine,* no. 10 (November 1993), 15.

238 **reduced or amputated the clitoris:** Cheryl Chase, "'Cultural Practice' or 'Reconstructive Surgery'? U.S. Genital Cutting, the Intersex Movement, and Medical Double Standards," *Genital Cutting and Transnational Sisterhood* 126: 145–146; Brian D. Earp et al., "The Need for a Unified Ethical Stance on Child Genital Cutting," in *Nursing Ethics* 28, no. 6–7 (2021): 1294–1305.

239 **a boy who had suffered a botched circumcision:** John Colapinto, *As Nature Made Him: The Boy Who Was Raised as a Girl* (Toronto: HarperCollins Publishers, 2000), 10.

240 **the vast majority of them medically unnecessary:** Nancy Ehrenreich and Mark Barr, "Intersex Surgery, Female Genital Cutting, and the Selective Condemnation of 'Cultural Practices,'" *Harvard Civil Rights-Civil Liberties Law Review* 71 (March 2005), 74.

240 **the first American hospital:** Thomas Buckley, "A Changing of Sex by Surgery Begun at Johns Hopkins," *New York Times,* November 21, 1966.

240 **3,000 letters from all over America:** Kaitlyn Pacheco, "Ever Forward," *Baltimore Magazine,* June 2019.

240 **just thirty-two of them had been at Hopkins:** Jane E. Brody, "500 in the U. S. Change Sex in Six Years with Surgery," *New York Times,* November 20, 1972.

241 **three were already married:** Stuart H. Loory, "Surgery to Change Gender: The 'Transsexual'—A Case Study," *New York Times,* November 27, 1966.

241 **"It was so highly sexualized":** Rachel Witkin, "Hopkins Hospital: A History of Sex Reassignment," *The Johns Hopkins Newsletter,* May 1, 2021.

243 **"penile scrotal-flap technique":** Raval and Hodges, *Trinidad.*

244 **"You have to see these people":** S. J. Guffey, "Town Labeled 'Sex Change Capital of the World,'" *Associated Press,* January 27, 1985.

244 **"my transsexuals":** Dennis McClellan, "Dr. Stanley Biber, 82; World Renowned Sex-Change Surgeon," *Los Angeles Times,* January 22, 2006.

244 **"Movie stars, judges, mayors":** Margalit Fox, "Stanley H. Biber, 82, Surgeon Among First to Do Sex Changes, Dies," *New York Times,* January 21, 2006.

245 **"the cat's whiskers":** Bill Shaw, "The Sex-Change Doctor," *Dallas Morning News,* December 28, 1984.

245 **criticism for its flawed methods:** Jane E. Brody, "Benefits of Transsexual Surgery Disputed as Leading Hospital Halts the Procedure," *New York Times,* October 2, 1979.

245 **"Hopkins was fundamentally cooperating":** Amy Ellis Nutt, "Long Shadow Cast by Psychiatrist on Transgender Issues Finally Recedes at Johns Hopkins," *Washington Post,* April 5, 2017.

245 **He would later admit that he came to Hopkins:** Charalampos Siotos et al, "Origins of Gender Affirmation Surgery: The History of the First Gender Identity Clinic in the United States at Johns Hopkins," *Annals of Plastic Surgery* 83, no. 2 (August 2019): 133.

245 **provided they had the funds:** Eric Plemons, "A Capable Surgeon and a Willing Electrologist: Challenges to the Expansion of Transgender Surgical Care in the United States," *Medical Anthropology Quarterly* 33, no. 2 (2019): 282–301.

246 **60 percent of the world's gender affirmation operations:** McClellan, "Dr. Stanley Biber."

246 **"As long as my hand's steady":** Michael Haederle, "The Body Builder: For 25 Years, Dr. Stanley Biber—America's Dean of Sex-Change Operations—Has Been Correcting Nature's Miscues," *Los Angeles Times,* January 23, 1995.

248 **"Trinidad's transgender rock star":** Douglas Brown, "Trinidad's Transgender Rock Star," *Denver Post,* June 29, 2007.

Afterword

252 **"Wikipedia will give you some sense":** Bo Laurent, email to the author, September 2, 2020.

252 **Bo fell silent:** Elizabeth Weil, "What If It's (Sort of) a Boy and (Sort of) a Girl?" *New York Times,* September 24, 2006.

253 **"The thing that they had done to me":** Bo Laurent, interview by the author, October 24, 2020.

254 **"I can make a good female, but it's very hard to make a male":** Norman Atkins, "Dr. Elders' Medical History," *The New Yorker,* September 26, 1994, 45.

INDEX

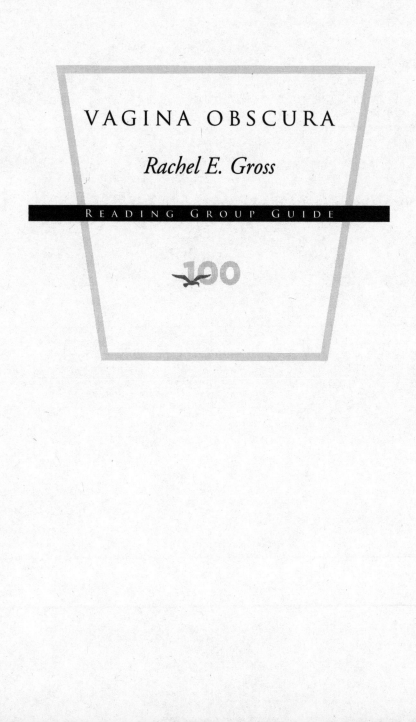

VAGINA OBSCURA

Rachel E. Gross

VAGINA OBSCURA

Rachel E. Gross

DISCUSSION QUESTIONS

1. *Vagina Obscura* drills into the background assumptions and biases that scientists—and all of us—have when it comes to female anatomy. What was one of the assumptions you had when you first opened this book? How did that understanding change by the end?

2. When did you first learn about the female sexual and reproductive system—the vulva, vagina, uterus, and ovaries? Do you remember feeling curiosity, shame, fear, or discomfort?

3. This book shows several ways language can shape our relationship to our bodies. When you think of the word "vagina," what are your first associations? What about "clitoris"?

4. Scientists and doctors who mapped the female body were often in pursuit of an elusive "normal." We see examples of this when it comes to the female orgasm, the vaginal microbiome, and the size of the clitoris. What are the dangers of designating one kind of body as the norm? Who do these standards end up hurting most?

5. Dr. Linda Griffith and Dr. Helen O'Connell are examples of scientists who have been able to imagine new scientific possibilities and challenge existing modes of thinking. How do the paths that led to their realizations differ? Are they models who can be followed to break old patterns of thinking? Why or why not?

6. Do you agree with Dr. Griffith's intention to avoid marking her Center for Gynepathology Research with traditional symbols of femininity? Why, or why not? Does this promote a gender-neutral approach to science, or does it perpetuate sexist notions about the role of women in science?

7. What is something you learned about human anatomy in this book that you found particularly interesting, inspiring, or shocking? Explain your thinking.

8. Were you ever uncomfortable while reading this book? When? Why?

9. Rachel E. Gross writes, "The beliefs we share as a society about sex and gender harm all bodies. Culture, and medicine, shapes bodies" (pp. 253–54). Do you relate to these statements? Is it possible to see ourselves and our bodies outside the influence of culture, medicine, and societally held ideas?

10. Given our modern understanding of human biology and social structures, how would you describe what Marie Bonaparte experienced as "frigidity"? In your opinion, is there a contemporary idea or prejudice analogous to the nineteenth century's "frigidity"?

11. Aminata Soumare describes feeling lighter after discovering that she has options in response to the genital cutting she was subjected to as a baby (p. 42). How is this similar to Bo Laurent's realization that "she could change her narrative" (p. 253)? Can you think of a time when you realized you had more options than you previously thought? What led to this realization?

12. Science, and its claim to objectivity, is a major theme in this book. Is it possible for science to be objective, given all the biases and injustices it has perpetuated? Are there any examples in the book of scientists who challenge the objectivity approach and provide a different path forward?

13. Do the illustrations in the book add to your reading experience? Did one stand out to you? Did any of the artistic choices or symbolism surprise you?

14. Was there any topic you expected to be covered in this book that wasn't (pregnancy, PCOS, the hymen)? Why do you think these topics weren't covered?

VAGINA OBSCURA

Rachel E. Gross

A CONVERSATION BETWEEN THE AUTHOR AND
ANGELA GARBES, AUTHOR OF *LIKE A MOTHER*

AG: I love that we're meeting to have this conversation. But in some ways I feel that, reading each other's books, we're already in conversation. We each wrote the book we wanted to exist, doing our best with the resources we had and the research that we did. But we both also wondered if our books were incomplete. The topic of female reproductive health is so vast. There is no definitive book, and we could never set out to write the definitive book.

So my question to you is: What is the book that you set out to write? What do you think made this a book only you could write?

RG: My mom is a doctor, and my parents are all scientists. I grew up loving to talk about animal sex and gross science and eventually, reproductive biology. When I started working at some of these legacy publications like *Smithsonian* magazine, I realized I was coming up against this culture that wasn't quite ready for that conversation. That there wasn't a space yet where these kinds of frank conversations could be had with a sense of curiosity, fun, and wonder—which I thought was the obvious approach to these things.

I also grew up without a particularly repressive religious structure. And so I had a very different approach than any other person might have to this topic. When I started writing, it became clear that shame was this really prevalent theme in the history of how anatomists mapped the female body and how they approached each of these organs, and that it was something I really couldn't escape. From Freud to Darwin, I realized that I was going to have to address this male-centric idea that female organs were for male pleasure, for reproduction, or were something dirty that should be kept hidden and secret.

And so I think I set out to reveal the roots of this shame and to show that they were cultural and historical, and manufactured. They had nothing to do with the actual science and reality of our bodies, which turned out to be far more expansive and resilient than I even imagined.

AG: I didn't have the upbringing that you did, but I feel like we arrived in the same place. I'm the daughter of Catholic immigrants from the Philippines, and I grew up with a profound amount of—not exactly shame, but silence. My parents are healthcare workers. My father is a physician, my mother is a nurse. We talked about bodies very clinically. I knew about frozen sections, I knew about autopsies. These kinds of terms were always thrown around. But in terms of the emotional, personal sides of bodies, there was so much silence. And I really feel like silence is where shame flourishes.

What both of us do in our books is recognize that shame is a problem. This idea of taking on Darwin and Freud, of taking on institutions to say this is not a female problem, this is an institutional problem, this is a systemic problem—that's a step that I think a lot of people aren't willing to make. It takes a certain amount of bravery. It's something that I really admire about your work: you name the structural forces, because everything about our society wants us to turn away from that and maintain the status quo. What gives you momentum?

RG: As I was doing archival research, I started to realize that the people coming up with these theories about women's bodies generally had little experience with women—and seemed to have little empathy for them. That was disturbing, and I wanted to correct the record. But I was also seeing these patterns that came up again and again that really pointed toward the fact that this was a systemic, institutional problem.

For instance, with Miriam Menkin, the IVF trailblazer, it was so clear the ways in which the system pushed her down at every turn while making her feel that her problems, like trying to become a scientist in the 1920s, were her own failings. Or when I talked to endometriosis patients, and I found myself hearing the same narrative over and over again, I started to understand that this couldn't possibly be a personal issue. The data just kept piling up to the point where I realized I had plenty of evidence to build this argument and to show that the problem was not rooted in any one person.

One thing I thought your book [*Like a Mother*] did so well was to illuminate how, the further you are out from this kind of white, feminine reproductive ideal, the more you are shamed or alienated or left out of the discourse. Setting that ideal itself was prescriptive. So my project became revealing those kinds of biases and forces that became this powerful, overarching paradigm about what it means to be a woman and what it means to have these body parts and what it means to be female.

AG: I love that each chapter is dedicated to the clitoris, the external, the internal, the uterus, the egg cell, the ovary. I wanted to ask how you decided on that structure, and specifically the bookends. I was really moved by the decision to begin with desire. That's a strong statement to start the book.

RG: Pleasure, to me, is absolutely central. So the book does start with a chapter—two, actually—on the clitoris. But in a way, it also ends with pleasure, because the chapter on gender affirmation surgery is about how this medical procedure ended up transforming from something that was about fitting a person into a strict gender binary to one that centers the patient, their wishes, and their desires, including pleasure.

It was very important for me to start the book from a place of curiosity and wonder. And I'm a sci-fi nerd. So I always had in my mind this kind of journey to the center of the earth structure, a kind of *Magic School Bus* journey into the body from the outside in. I also wanted to make sure to surprise with each organ, to make you see your body in a different light. So we would zoom in on the egg cell and then zoom out to the entire ovary, and by the end you would feel like you'd gotten to see all these different facets of the body in a way that you never had in sex ed or biology class.

As I was researching, I realized that each of the organs I was looking at was conceptualized in this very specific way, usually by the men who first explored and planted their flag in them, and that now a new generation of researchers was reimagining them. I wanted to capture that project of reimagination, to move us forward. To ask what it would mean to take seriously the idea that the uterus is also a regenerative organ even when it's not pregnant, and that the ovaries could make new eggs. And how that could point us toward a new frontier of science.

AG: One of the things I really love about this book is that, yes, it's about uteri, it's about ovaries, it's about magnificent eggs, it's about vaginas and clitorises. But it's really a story of people. And you realize that we learn these things because someone was committed to studying them, despite all the things that were going on in their lives. I'm thinking specifically about Helen O'Connell, Marci Bowers, and Miriam Menkin. I feel the gratitude that you have for them and a sort of admiration. What is your relationship to these sources? And did you come to ground much of this book in their journeys?

RG: Right, and it wasn't just despite their journeys that they came to commit to this; it was often *because* they had felt those oppressive forces. They had seen that they were not part of the mainstream medical narrative, and had felt that they needed to right this wrong, fill in this gap, and expand the lens to include people like them. It was that productive, loving critique: instead of just saying this field is broken and there's no place for me here, they said, "I'm going to fight until we figure out what's going on and until we balance the scales."

Before I wrote this book, I'd been running this column on unsung women in the history of science for a few years at *Smithsonian* magazine and then BBC Future, with this brilliant historian-writer Leila McNeill. Each one was an in-depth profile that was half about this woman's individual life and her struggle to transform her field, and half about the systemic forces that worked against her. In this case, getting to talk to people who were doing this in real time—not just women—and who oftentimes were still struggling to be heard was such a privilege.

As they told me their stories, I was, on the one hand trying to distill the science and show how their work transforms our understanding of the female body. But I was also hearing so many unscientific barriers that they were up against, and background sexism, like being the first female urologist in an entire country and realizing that no one else had asked, "What about the nerves of the clitoris during surgeries?" To me, those moments revealed something universal about what the people who tried to do this work were facing. These are challenges that the next generation of explorers will undoubtedly come up against.

There's a theme here. There's this project of documenting the patterns of how these systems make you feel isolated and alone, like you're the only one swimming upstream. But really, I found there is also a community here. And quite remarkably, some of the scientists in this book have now talked to each other, through Twitter or conferences or even by reading this book. I do think that there's a kinship there. Not just the potential for scientific collaboration, but an understanding that they were part of a bigger project that they didn't necessarily know about.

Linda Griffith once said to me that this work is a big mosaic, and none of us can see the entire picture. What we can do is put down our own tile and start to fill that in, and slowly, the picture emerges. She was talking about scientists, doctors, patients, and me, which was both really gratifying and embarrassing. But I do think that it's true. I couldn't see the tile I was filling in when I started this book. But now, I think, I can.

AG: Change comes from multiple ends, and we need everyone involved to accomplish change. I think about how we are each experts when it comes to our own lived experiences and bodies. I don't think the medical establishment likes to hear that, and I think people are uncomfortable hearing that. But I believe that's true, and if someone reads something we write and feels compelled to ask a question as a patient, well, I want to encourage people to speak into that more.

RG: This is something that I strongly saw when I started talking to patients who had endometriosis or other reproductive diseases.

Many of them knew for years that something was deeply wrong inside them. They may not have had the language to articulate it, but they had this embodied knowledge that they could not communicate to their doctor or that their doctor could not hear. The medical profession needs to find a way to bridge that language gap or that authority gap—both for the sake of patients, and because these patients and their stories have so much to offer medical knowledge.

AG: I'd love to hear you talk about interrogating, queering, the science behind the binary, and how you square that with celebrating these body parts that are so deeply associated with womanhood.

RG: As science journalists we know that clarity is paramount. One of the worst things I can do is lose someone on the first page because my language is too confusing or technical. In that case, the reader doesn't even get to the complicating / queering part. I do use "woman" as shorthand quite a lot, and I try to explain my reasoning up front. I try to get that over with so that I can take you through the history of when people with these body parts were generally lumped into the same category, and show you how profoundly those categories are changing.

But there's no way around the fact that you're not going to include everyone while also being perfectly specific and accurate. You have to do the best you can and build the widest net you can while being clear. The way to expand the lens and ultimately make science less biased is to bring in all the voices that haven't been heard. And so I invited into this book a lot of people who had backgrounds and lived experiences that I didn't. And often the result was very eye opening.

One person who really shifted my paradigm was Joan Roughgarden, a biologist and transgender woman who made it abundantly clear to me that biologists today are still very heterosexist, and that Darwin's shadow of thinking that females are chaste and passive and only exist in monogamous pairs still dominates the field. She's one of the reasons we are opening up a new field of queer biology. I had not been aware of those biases. I had probably been playing into them when I told the story of duck vaginas and Patty Brennan, who as she herself admits is more focused on sexual conflict and heterosexual mating.

So that was when I realized I had this deep-seated bias that I needed to unlearn. In the book I tried to capture those basic realizations and revelations that I was going through, and I'm so grateful for the people who helped me do that. I couldn't have done that myself. I feel like their fingerprints are all over the book.

AG: I think it's important to be humble while you're writing and to be schooled a little bit. Sometimes it's harsh, but that's how you get a better book. You don't know what you don't know, you don't see what you don't see until you invite someone in. That is your work as a journalist: to invite all of the different perspectives, to have those fingerprints and to not be threatened by that. This myth of objectivity forces a lot of people into getting defensive, when actually it's to your benefit to be porous.

RG: Yes. And often these are the people who are pushing the boundaries of what counts as science, and what science needs to look at next. They're drawing out these important connections between all bodies. For instance, in the last chapter, I learned that for medicine to truly create a new vagina, researchers have to go back to the basics and truly understand what the vagina is and how it does all of the amazing things it does—which we don't, yet. We don't know how the vagina self-lubricates, we don't fully understand the vaginal microbiome.

I remember when Patty, the duck vagina researcher, made this point where she basically said, "Yeah, I do talk about vaginas and clitorises a lot. Guess what? For most of history we haven't. And so it's going to be all about the vagina for a while. We're going to have to make up some ground here."

There's something to that. The making of new knowledge and going against existing paradigms is going to have its own missteps and oversights, and science will have to correct for that. But for those of us who have looked into this history, we know exactly how unbalanced and dismissive science has been of these organs and their scientific potential. And I think chronicling how far we've come, and the mere fact that what everyone thought the female body was capable of was probably wrong for hundreds of years, is deeply important.

So yeah, I'm okay with overshooting a little bit, if it means changing the lens to focus on the vagina.

AG: Right, it's not like you're thinking, I'm learning about the vagina so that this can be absolute knowledge. You're not saying this is the end, the goal. You're arguing that the goal is to keep going.

RG: That's the thing about a book. There's something about it that's supposed to be finite and finished and definitive. While I was working on the book, I actually had a Post-it on my desk that said: a book is a moment in time. The science will change, and that's okay. You can't incorporate every new study. And people will understand that this is the state of science as it was—as it is—now. It took a lot to come to terms and be okay with that. We're both just starting this conversation. I want scientists and science journalists to follow up on the new science that is still unfolding, and come back and tell us what's going on in the ovaries, and exactly how resilient and dynamic these organs really are.

The other thing to keep in mind is that we're meeting readers where they are. There's always a balance: How do you welcome the most people in, and how do you speak to the people who are most educated or the most ahead of the curve? I found that a lot of the things I worried were common knowledge, like the size of the clitoris, are still the things that people tell me really made them rethink their bodies. Everybody needs a window into this subject matter. And we, the writers, can provide that window. We serve a function. And that is not to be the omniscient, objective, transparent eyeball.

AG: One of the things I love about the book is the wonderful way it draws on the natural world and other animals. To me that is such a powerful reframe. When I read about the duck vaginas, it resonated with me so deeply in this political moment as a way of having reproductive freedom. The idea is that you can't always stop what happens to your body, but you can choose how—or if—you reproduce. There's power and agency in these stories from the natural world.

RG: We're in a political moment where there's a lot of political efforts to stamp out variation, whether it's trans folks or gay marriage or trying to dictate what women's bodies are for. And there's a strong strain of what is natural at the base of this, as there always has been. What I've found is what is natural generally works to serve those in power. There is something subversive in going to the natural world and saying, "What is nature actually doing and what is natural variation?"

To me, the danger here is saying that any one animal is superior or the "goal." I think Joan put it best when she says nature gives us so many examples of how to live, and how to have a body, and how to do sexuality and reproduction. And the beauty of being human is that we can pick and choose what speaks to us and what we identify with, just as you had your own reaction to the duck vagina.

Variation is literally the material by which nature moves forward and evolves. It is the reason that we exist today. And to stamp it out just seems so shortsighted and so unappreciative of that expansive beauty. Nature doesn't have one prescriptive way for how to be human, or how to be female. It all goes back to the theme of the book, that the female body, whatever that means to you, is more resilient and dynamic than it's been given credit for. That the organs we think of as reproductive are really doing so much more.